Managing Factory Maintenance

SECOND EDITION

Joel Levitt

Springfield Resources,
401 Rolling Hill Dr.
Plymouth Meeting, Pa. 19462
Voice: 610-278-7550
fax: 610-278-7552

JDL@Maintrainer.com

Industrial Press Inc.
New York

Library of Congress Cataloging-in-Publication Data
Levitt,Joel, 1952-
Managing factory maintenance / by Joel Levitt.--2nd ed.
p. cm
ISBN 0-8311-3189-6
1. Plant maintenance--Management. I. Title
TS192.L47 2004
658.2'02--dc22

2004017939

Managing Factory Maintenance
Second Edition 2005

Industrial Press Inc.
200 Madison Avenue
New York, New York 10016

Text and Cover Design: Janet Romano
Manageing Editor: John Carleo

Disclaimer: The advice given in this book is general. The judgment of qualified people is essential to determine the fitness of any of these statements for your use. All statements must be adapted to the site conditions of any individual shutdown. The author and publisher take no responsibility for any adverse outcomes.

10 9 8 7 6 5 4 3 2 1

Dedication

I have been blessed with the opportunity to teach maintenance management topics to thousands of maintenance professionals worldwide for universities, private companies, publishers, and soft-ware companies. I want to dedicate this work to those folks who sponsor professional development training.

The idea for this work originally came about from a request to give a course in Shutdowns and Outages in Trinidad from Anne DeSousa of Trainmar. Thank you Anne and your able assistants Camille Whitby, Janna Turpin and Neisha Mohammed. The course subsequently was moved to South Asia, sponsored by Partners in Singapore. Thank you Delphine Ang and Rona Lim.

The able folks at New Standard Institute helped in that Mike Brown taught me his approach to shutdowns and planning and Tessa Marquis got me the opportunities to teach. Thank you both.

I have been giving courses for the University of Alabama since 1989 and have enjoyed the connection with this great institution. Thanks to Tom Wingenter and his able staff for running an excellent program. My first sponsor was Clemson University and USC (now called Saddle Island Corporation). The person who observed me in 1987 and thought I'd make a decent instructor was Jean Hey. Thanks wherever you are. I gave over 100 courses for Betsy Tyson at Saddle Island, Thanks.

There were many others in the past that worked diligently to produce good and educationally sound programs on maintenance management. I appreciated the opportunity to represent you and provide the best training that I could in these topics.

Sincerely,
Joel Levitt

Table of Contents

MAINTENANCE INTERFACES: WHERE DOES MAINTENANCE FIT IN?

PERSONAL AND PERSONNEL DEVELOPMENT

How the New Revised Edition is Organized

There have been several changes in the field of managing factory maintenance since 1996 when the first edition came out. This revised edition has been extensively rewritten and reorganized to reflect the changes.

The two biggest changes are the globalization of the location of production and the changing of the guard of the leadership of maintenance. Each trend was clearly present in 1996 but neither had become the 800-pound gorilla that they are today (they were merely 400-pound gorillas back then). These two monumental shifts present peril and (as always) tremendous opportunity.

The book is organized to more closely reflect how it is to be used. The book can be read as a coherent story about managing factory maintenance but it is also designed for a person interested in or responsible for maintenance to get relatively specific answers to any big questions they might ask. The chapters are now presented in a context of major sections that tell the story of maintenance management in factory settings.

Section: **What is the Context for Managing Maintenance?** *Why we are in this situation:* What happened to produce the situation we are now in. What potential problems and new opportunities result from the present situation? *What bosses really want:* There is a lot of noise from management. What do they really want from the maintenance effort? When the comments are examined it is found that top management is really asking for something pretty simple. Management wants to run the factory with 100% availability, safely, and in compliance with the law, without any maintenance department at all!

Specifics to improvement of maintenance and production: To achieve that goal we must increase our effectiveness, reduce waste, and do more with fewer people.

World Class Maintenance Management visited and revisited. I've thought about world class maintenance for the last 8 years. Here is the original view and some revised thoughts.

Section: **Evaluating current maintenance practices.** Many questions have changed, moved or been eliminated from the first edition. Maintenance processes have not really changed in 8 years. What has changed is the enabling technologies. What has also changed is the deep proliferation of systems in maintenance departments. As with every other area of the business, computer expertise is moving rapidly to the masses from the IT departments.

Section: **Maintenance processes:** Description of the processes to deliver maintenance services. This section has been reorganized to suit new techniques and technologies.

Quality improvement: We learned to importance of quality in manufacturing in the 1980s and 1990s. Quality has had a chance to merge into the culture now.

Section: **Maintenance process aids:** Formally we thought that the initials represented maintenance management in themselves. If we had PM, TPM, RCM, or PMO we thought we were doing maintenance management. Actually these initials represent process aids. The underlying processes must be whole and complete for the aids to help us.
Section: Maintenance strategies: Approaches to deterioration: Strategies must fit the circumstances. To the degree that we remember this axiom we will be successful in organizing and improving the maintenance effort.

Section: **Maintenance interfaces:** Where does maintenance fit in? Maintenance has interfaces to all other parts of the organization. In this section, we discuss effective ways of interacting with our colleagues.

***Section*:** **Personal development and personnel development:** Whether we like it or not, we are in the training and development business. This section is designed to help the leadership organize and regularize training and development.

List of Figures

Introduction to the Second Edition

Why we are in this situation in the first place

Managing the need for maintenance is essential for any organization to survive in manufacturing .. Of course, a few companies (like a notable one in the steel business) have abandoned manufacturing because they feel they can make more money in another field. The mission of this work is to identify the challenges and opportunities presented by staying in the manufacturing field.

Our greatest challenge is two pronged.

Competition: A manufacturing manager was complaining that low wages were causing jobs to go off shore. He wanted the government to help him compete with the foreigners. This conversation could have been happening in Detroit, Rochester, or Omaha. In fact, the conversation happened in Kuala Lumpur, the capital of Malaysia. The jobs in question were moving from factories in Malaysia to new factories in Vietnam. In the next few years, it is expected that Vietnam will be a low wage-manufacturing powerhouse.

The first prong of the challenge is the increasing globalization of the location of production. Corporate management is seeking the lowest cost of production irrespective of other factors. In the US, we complain about the low cost of foreign labor but in fact, this is a challenge for all manufacturing centers worldwide. The low cost producers of today have to look out for the lowest cost producers of tomorrow.

In the US, manufacturing jobs have been moving to southern states for 40 years. It took 25 years or more to move production from domestic factories to plants first in Mexico then in Asia. The process is speeding up. It took only a few years for Korea, Taiwan and other powerhouse manufacturing centers to realize they could also save money by going to China or Vietnam. Where process innovation to offset advantages in labor cost is not taking place you can count on manufacturing seeking the lowest labor cost.

Loss of old timers with their skills due to retirement and retrenchment: The second prong of the challenge is due to demographics, business reorganization, and rapid changes in technology. Industrialized countries are losing their most experienced workers at an extremely rapid rate. By rapid we mean, for example, a power plant in the western United States will lose 66% of its maintenance workforce to retirement in the next 5 years!

The demographics are worse in Japan and Europe. In the US the higher rate of immigration is partially shielding us from the effects of the aging of the population and the lowered birth rate.

The trend is particularly noticeable in the maintenance shop because of the difficulty of improving productivity to adjust for fewer people. In the late 1990s, companies downsized the maintenance ranks to improve profits, and the hardest hit plants have suffered increasing deterioration. For the last few years, there has been a resurgence in hiring maintenance workers and an increasing difficulty in finding suitable candidates to replace the retirees. This trend is not unique to maintenance by any means but increases in productivity have partially covered up the same trend in the production ranks.

Loss of confidence in new technology is another major reason for some of the early retirements. One example occurred in a small recorder manufacturer in the UK where the chief of field service was 59 years old. The workforce consisted of three younger technicians. He was required to do service himself and usually took (or was brought) the most challenging problems.

He was the best and most highly skilled trouble-shooter for over 20 years with that company, his career having started with relay logic.

At some point microprocessors started to show up in his firm's equipment designs. He started to learn about them, but he was never given formal training. He never understood the concept of programming or the dynamic nature of the data and address bus. He could not understand how to troubleshoot a dynamic system such as a typical microprocessor board.

This dedicated field service manager took early retirement, feeling that the world had passed him by. The company lost his expertise. His knowledge of field conditions and problem solving was still unsurpassed, but he lost his sense of mastery and feeling that he was part of an important field. He now is the local electrician doing odd jobs in his neighborhood. Interestingly, his old subordinates still bring him problems that he throws himself at with relish.

Thus, the greatest local challenge faced by factories is the retirement of their old-timers (for whatever reasons). Some of these people built the plant and stayed on to operate it. They have specific plant knowledge that is not replicated in any database or on any drawing.

The peak retirement period started a few years ago and will continue through the rest of this decade. It is essential to learn from these people as they leave. The old-timers can provide stability for the manufacturing process from their vast storehouse of knowledge.

Taken together, the two challenges put us on the edge of a mountain ridge with thousands of feet to fall on both sides. To make the picture more accurate, the fall on one side is precipitous. Your plant can be shut down with the inking of an outsourcing agreement. On this side, one false step and you will fall thousands of feet to instant death in the valley below. The fall on the other side is gentler and will take longer. In a few short years, 80% of the years of maintenance expertise will be gone. Make no mistake about it, you will be dead in both instances.

Without the old-timers, how do we compete? The retirement of these experienced people paves the way to destruction from lack of stability of our processes (falling on the gentle side of the ridge). Destruction is gradual from competition gnawing away at market share. The other alternative is destruction from outsourcing the product made in your plant to a lower labor cost area (the precipitous plunge that shutters the plant overnight). There is a third course. It is perilous but we are facing a slow death or a quick death otherwise.

Although this situation is a great time of challenge, it is also a great time of opportunity. This knowledge, combined with the years of experience of the old timers can act as an anchor and slow the ship of innovation. We can use this opportunity to break through the way we have always done things to a new standard. We can pursue opportunities and efficiencies that would have been impossible with the old-timers in control.

To travel the ridge trail we have to get as much knowledge from the old-timers as possible to keep our process stable, and we must simultaneously look to new and innovative ways to produce our products. The level of innovation will have to be unprecedented to offset both the labor cost advantage and the innovations of our competitors.

The ridge trail is like the Chinese curse "may you live in interesting times." The perilous trail we face is not for the weak hearted. If you entered a maintenance department in 1960 and retired in 1985 you thought things were changing rapidly. If you entered the field, in1990 just wait to see how fast things will change in the next few decades.

Selected Glossary of Maintenance Management Terms

Autonomous Maintenance: Routine maintenance and PM's are carried out by operators in independent groups. These groups solve problems without management intervention. The maintenance department is called on to solve bigger problems that require more resources, technology or downtime.

Asset: May be a machine, a building or a system. It is the basic unit of maintenance and the driver of the PM and computerized maintenance systems.

Backlog: All work for the maintenance department that has been formally identified with a work order. While it is in Backlog, identified work is approved, parts are either stocked or bought, and (in ready backlog) everything is ready to go.

BNF equipment: Equipment left out of the PM system, left in the Bust 'N Fix mode (it busts and you fix it - no PM at all). BNF is a choice used when PM is not advantageous.

Capital spares: Usually large, expense, long lead-time parts that are capitalized (not expensed) on the books and depreciated. They are kept for protection against excessive downtime costs.

Call back: A job for which a maintenance person is called back because the asset broke again or the job wasn't finished the first time. See rework.

Charge rate: This is the rate that you charge for a mechanic's time. In addition to the direct wages you must add benefits and overhead (such as supervision, clerical support, shop tools, truck expenses, and supplies). You might pay a tradesperson $25.00/hr and use a $65/hr or greater, charge rate. It is important to compare your true cost of doing a large maintenance job with the cost of using a contractor.

Continuous Improvement (in maintenance): Reduction of the inputs (hours, materials, management time) to maintenance to provide a given level of service. Can also mean increases in the number of assets, or use of assets with fixed or decreasing inputs.

Core damage: When a normally rebuildable component is damaged so badly that it cannot be repaired.

Corrective maintenance (CM): Maintenance activity that restores an asset to a preserved condition. CM is normally initiated as a result of a scheduled PM or PdM inspection. See also planned work.

Deferred maintenance: All the work you know needs to be done that you choose not to do. You put it off, usually in hopes of retiring the asset or getting authorization to do a major job that will include the deferred items. You worry that the asset will fail before you get to it.

DIN work: 'Do It Now' is non-emergency work that you have to do now. An example would be moving furniture in the executive wing.

Emergency work or emergent work: Maintenance work requiring immediate response from the maintenance staff. Emergent work also refers to work that emerges after you open up an asset (pump, vessel, etc). Emergency work is usually associated with some kind of danger, safety, damage, downtime, or major production problems.

Frequency of Inspection: How often do you do the inspections? What criteria do you use to initiate the inspection? See PM clock.

Future Benefit PM: PM task lists that are initiated by a breakdown rather than a normal schedule. The PM is done on a whole machine, assembly line, or process, after a section or sub-section breaks down. This is a popular method with manufacturing cells where the individual machines are closely coupled. When one machine breaks, the whole cell is PM'ed.

GLO (Generalized Learning Objective): The general items necessary to know to be successful in a job. Each job description would be made up of a series of GLO's.

Iatrogenic: (Formal definition: Induced inadvertently in a patient by a physician's activity, manner, or therapy.) In our case it describes failures that are caused by your service person.

In-box: In-box jobs are the first step of the maintenance work order process before the job is reviewed and entered into backlog.

Inspectors: The special crew or special role that has primary responsibility for PMs and PdMs. Inspectors can be members of the maintenance department or of any related department (machine operators, calibration, drivers, security officers, custodians, etc.)

Interruptive (task): Any PM task that interrupts the normal operation of a machine, system or asset, and requires maintenance to take custody of the asset.

Life Cycle: Denotes the stage in life of the asset. Three life cycles or stages are recognized by the author: start-up, wealth, and breakdown.

Life cycle cost (LCC): A total of all costs through all the life cycles. Costs should include Maintenance (PM, repair including labor, parts, supplies, outside services, rentals), downtime, energy, ownership, overhead. An adjustment can be made for the time value of money.

MTBF (Mean Time Between Failures): Important calculation to help set-up PCR schedules and to determine reliability of a system.

MTTR (Mean Time to Repair): This calculation helps determine the cost of a typical failure. It also can be used to track skill levels, training effectiveness, and effectiveness of maintenance improvements.

Management: The act of controlling or coping with any eventuality.

Maintainability Improvement: Also Maintenance Improvement. This activity looks at the root cause(s) of breakdowns and maintenance problems and designs a repair that prevents breakdowns in the future. I Includes improvements to make the equipment easier to maintain. .

Maintenance: The dictionary definition is "the act of holding or keeping in a preserved state." The dictionary doesn't say anything about repairs. It presumes that we are acting in such a way as to avoid the failure by preserving the asset.

Non-interruptive task: PM task that can safely be done without interrupting production by the machine.

Non-Scheduled work: Work that you didn't plan for and didn't schedule.

PCR: Planned Component Replacement. Component replacement to a schedule based on MTBF, downtime costs, and other factors.

Parts: All the supplies, machine parts, and materials, needed to repair an asset, or a system in or around an asset. In some cases parts are separated from supplies and consumables.

Planned maintenance: Maintenance jobs for which all resources have been identified. Once the resources have been written into the planned job package, the job can be scheduled for execution. See also scheduled work

PM: Preventive Maintenance is a series of tasks that either, extend the life of an asset or detect that an asset is at a critical point and is going to fail or break down.

PM Clock: The parameter that initiates the PM task list for scheduling. Usually buildings and assets in regular use, days (For example, PM every 90 days) by the clock. Assets used irregularly may use other production measures such as pieces, machine hours, or cycles.

PM frequency: How often the PM task list will be done. The frequency is driven by the PM clock. See frequency of inspection.

Predictive Maintenance: Maintenance techniques that inspect an asset to predict when a failure will occur. For example, an infrared survey of an electrical distribution system might look for hot spots (which would be likely to fail). Predictive maintenance is usually associated with advanced technology such as infrared or vibration analysis.

Priority: The relative importance of the job. A safety problem would be more important and be fixed before an energy improvement job.

Proactive: Action before a stimulus (Antonym: reactive). A proactive maintenance department takes actions before a breakdown occurs.

RCM: Reliability-centered Maintenance. A maintenance strategy designed to uncover the causes and consequences of breakdowns. RCM sets up priorities by the severity of the consequences. PM tasks and redesign are directed specifically at those failure modes that have the worst consequences. RCM is a procedure for uncovering and overcoming important failures.

Reason for write-up (also called reason for repair): Why the work order was initiated. Reasons could include PM activity, capital improvements, breakdown, vandalism, and any others peculiar to the industry concerned.

Rework: All work that has to be done over. Rework is bad and indicates a problem with materials, skills, or scope of the original job. See also call back.

Root cause (and root cause analysis): The root cause is the underlying cause of a problem. For example you can snake out an old cast or galvanized sewer line every month and never be confident that it will stay open. The root cause is the hardened buildup inside the pipes. Analysis would study the slow drainage problem and figure out what was wrong. The study would also estimate the cost of leaving the defective pipe in place. Some problems (not usually this type of example) should not be fixed.

Route maintenance: Mechanics who have an established route through your facility to fix all the little problems reported to them. The route mechanics are usually very well equipped so they can deal with most small problems. Route maintenance is sometimes combined with PM activity.

Routine work: Work that is done on a routine basis where the work and material content are well known and understood. An example is daily line start-ups.

SLO: Specific Learning Objective. The detailed knowledge, skill, or attitude, necessary to be able to do a job.

Scheduled work: Work that has been planned and is written on a schedule at least a few days in advance. Many writers use the phrase "planned maintenance" to refer to maintenance work that is both planned and scheduled.

Short Repairs: Repairs that a PM or route person can do in a short time with the tools and materials that they carry. A short repair is a complete repair that can be done in a short time. It is different from a temporary repair.

String based PM: Usually simple PM tasks that are strung together on several machines. Examples of string PM's would include lubrication, filter change, or vibration routes.

Survey: A formal look around at the overall condition of all assets, when all the aspects of the facility are recorded and defined. The survey will look at every machine, room, and through-out the grounds. The surveyor will note any thing that appears to need work.

TPM: Total Productive Maintenance. TPM is a maintenance system set-up to eliminate all of the barriers to and losses in production. TPM identifies production losses and uses production operator teams to solve the problems causing the waste. Autonomous maintenance teams (focusing on operators) are used to carry out most basic maintenance activity.

Technical Library (Maintenance Technical Library): The repository of all maintenance information including (but only limited by your creativity and space) maintenance manuals, drawings, old notes on the asset, repair history, vendor catalogs, MSDS, PM information, engineering books, shop manuals, and so on. The maintenance technical library could be virtual (stored entirely on computer).

Task: One line on a task list (see below) that gives the inspector specific instructions to do one thing.

Task List: Contains specific directions to the inspector about what to look for during that inspection. Typical tasks could be inspect, clean, tighten, adjust, lubricate, replace, etc.

UM: User Maintenance. Any maintenance request primarily driven by a user. Includes small projects, routine requests, and DIN jobs.

Unit: See also asset. The unit can be a machine, a system, or even a component of a large machine.

Work Order: Written authorization to proceed with a repair or other activity to preserve an asset.

Work request: Formal request to have work done. Work requests are generally filled out by a maintenance user. Work requests are usually time/date stamped and form the basis of the work orders

What is the Context for Managing Maintenance?

What do bosses really want from the maintenance effort?

We don't have to be mind readers about what the big bosses want from maintenance. We just have to read the Wall Street Journal or any newspaper business section. Big bosses want less maintenance, big bosses want maintenance that does not interfere with production, and big bosses don't want anything like accidents, environmental violations, or fires, to get in the newspapers.

The bosses are responding to the reality of their market places. They don't necessarily see the retirement of skilled maintenance workers as a core issue but they do see the erosion of market share by competitors (both domestic and international). Bosses are constantly being exhorted by corporate management to lower the unit cost of production. In many companies, if the unit cost can't be lowered, production will be moved to lower labor rate areas overseas or to plants with lower overall costs.

These conditions are the reality of the ridge road. Slow death on one side from erosion of market share, and quick death on the other from a plant closure. The ridge road is a tough road because the maintenance department is smaller and there is less opportunity for mistakes. The consequences of any mistakes are greater.

How do we measure this effect?

The ideal plant is bigger (more output without additional assets) because productive machines in a plant may break, because machines are not run to nameplate speeds, and a variety of other reasons. Maintenance has an impact on many of these items and can positively impact the others through getting involved and lending its expertise.

The easiest way to see this effect is to visualize your factory with a size proportional to output.

This is your current factory (size proportional to output)

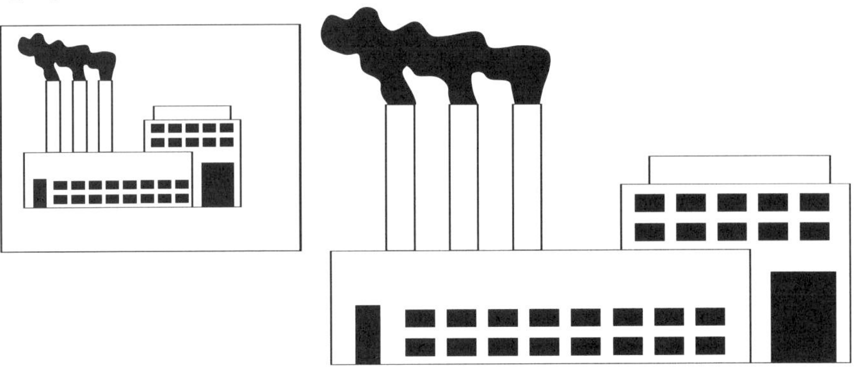

This is your ideal factory, or what you could produce if everything went right

One measure developed to evaluate factory output by the TPM folks is OEE. OEE (Overall Equipment Effectiveness), which is a measure of the amount of effective output compared with the ideal output possible from the same plant, area, or machine. A typical factory might have an OEE of 50% to 80%. The 20% to 50% that is left represents wasted resources. The waste comes from breakdowns, model changes, material problems, small jam-ups, etc. Without spending any money on expansion, most plants could increase their output by half of these numbers (10%-25%).

Although reduction in the cost of maintenance is an admirable goal (and will be dealt with extensively in this text), the real money is in increasing the OEE of the whole plant.

Everything you ever wanted to know about maintenance can be learned on Star Trek

Since 1967 Star Trek in its various forms has been a successful US TV series. It has undergone several redesigns. The maintenance message of the three main series is really all you need to know!

In the first series, the Chief Engineer was Montgomery Scott. He was a down and dirty maintenance guy from the old school. You would routinely see him crawling around the engine room with weird looking tools, fixing things. Scotty was a super repairperson with a complement of cool tools. Over time we find out that he is an accomplished engineer and designed the standards that all Star Fleet engineers use. Scotty was the 60s' vision of the ultimate maintenance guy. Scotty is paternal, tough, and competent.

In Star Trek, The Next Generation, the Chief Engineer is Geordi La Forge. Geordi is blind from birth but sees the entire Electrical Magnetic spectrum (as well as some other cool capacities) with his visor. In 100 episodes Geordi rarely, if ever, repairs anything. If there is a problem, he waltzes up to a computer console and reconfigures the Warp couplings (or whatever). He maintains the ship completely by computer! Occasionally when something strange happens and the computer fails he is also the ultimate repair guy, but this happens infrequently. We find out that he is also a leading physicist. He is the ultimate 90s maintenance guy using the computer to fix everything. So Geordi is hi-tech, personable, competent genius that is comfortable chatting up leading theoretical physicists and can also jump in and fix things

In the third series, Voyager the Chief engineer is B'Elanna Torres. Her ship was swept into the Delta quadrant (very far from home, it will take 70 years to get home, even at Warp 10) by Q (a childish omnipotent being). Her ship has some biology built in so it can repair itself. So unless they were attacked or run into some weird anomaly in space (which does seem to happen pretty often) the ship itself can fix most things. Ms Torres spends most of her time trying to coax a little more power from the Warp engines to get home faster. B'Elanna is the ultimate 2000's maintenance person, no longer in the repair business but in the business of increasing output. She is powerful, loyal, passionate, and competent but is focused on the productive output not the repairs or maintenance at all.

We in maintenance contribute to the success of the organization. Our efforts can place the organization squarely in the middle of an admittedly narrow path. The goal of maintenance (like Star Trek Voyager) is eventually to eliminate the need for maintenance departments! The goal of maintenance is to do everything in its power to increase the quantity and quality of production, and reduce costs

How to Improve Maintenance

Improvement- first of all what is the goal of the effort?

PM approaches have always been worrisome. If a proactive approach was so superior, why hasn't that way of doing business taken over?

Maintenance Improvement Graphically These curves represent the average life (or MTBF - mean time between failures) of a component such as air cylinder (or bearing, seal, etc) in different maintenance strategy scenarios. Three strategies are represented: do nothing (called bust'n'fix), PM, and permanent maintenance improvement.

In the 'Bust and Fix' scenario assets are allowed to break down naturally. Left alone, each cylinder will deteriorate and fail in a given amount of time (represented by curve A, at the extreme left). With greater numbers of cylinders the graph will look more and more like a normal distribution or bell shaped curve (our diagram is simplified because differing failure modes will make the actual distribution more complex).

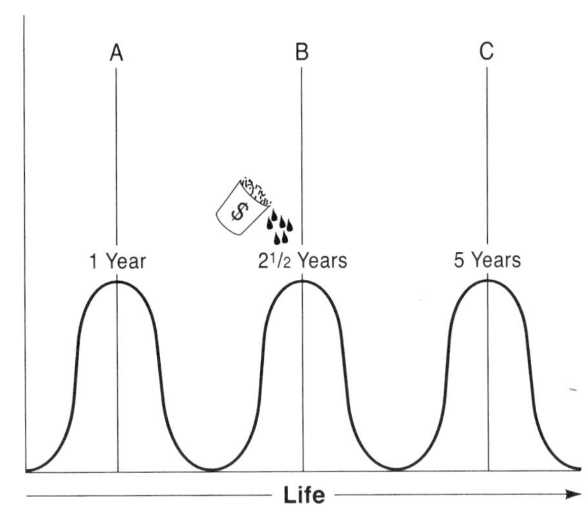

The relationship of PM to maintenance improvement

If you have more than one item, not all of them will fail at once. If you add up the whole area under the curve you would get the whole population of air cylinders (or machines). Eventually they all break. Think about cylinder failure as being like a bag of popcorn. Each kernel represents an air cylinder (bearing, seal, or machine). When you heat it (use it) some pop early (premature failures we describe as infantile mortality) some pop late, but the bulk pop in the middle of the process.

When they pop is represented by the position from the Y-axis and how many pop is represented by the distance from the X-axis. The curve might look like `A'. The MTBF will be influenced by how they are used, how well they are engineered and built, and the conditions under which they are operated.

Curve `B' represents an improved life resulting from PM and predictive maintenance. We are cleaning the cylinders, keeping everything tight, adding an appropriate amount of lubrication, and so on. As you will notice, the life curve has shifted significantly to the right showing longer average life. To keep the life at this level, funds have to be continuously invested in the form of PM labor and materials. As soon as that flow of money stops, the curve will slide back to `A'.

Herein lies the worrisome aspect of PM. When a new manager comes into a facility there is a temptation to want to shine or at least look good. If the manager chooses to cut back on PM, the chart shows that there will be no impact on failure rate (and therefore reliability) until enough time passes for the curve to decay to 'A'.

The profit will go up in the short term. The new manager will look like a hero for increasing profit. If he or she can get a new job assignment (with greater pay and privileges, of course) before the curve decays, they will leave as a hero and the next poor manager is stuck with the results of the bad decisions.

But what if we could impact the failure rate in a more permanent way? What if we could make progress and change the nature of failure in our factory, forever? Curve `C' is the goal of maintenance. It is called maintenance improvement, where the life of the unit unattended is longer then either `A or B'. This improvement could be the result of using better seal kits, better cylinders, and better protection. But we want to increase the MTBF without having to pour money in every month.

This improved ratio is the holy grail for maintenance, so this is where the attention should be focused. Of course, the maintenance improvement should be logical (we would not spend $50,000 to avoid a $100 problem). All improvement efforts in the maintenance department should have maintenance improvement as one of the points of interest.

Office of Continuous Improvement

To deliver what our higher-ups want we have to insure that each year we can produce the same products with less input or more products with the same input. We know that without continuous improvement in the delivery of maintenance there is stagnation, complacency and the fall from the ridge trail described earlier. What had seemed like a secure and stable situation in actuality is in constant flux and staying still is a prescription for disaster. This state is particularly true when management realizes that maintenance is not keeping pace (and everyone else is).

Research and development is a feature of all advanced organizations. They want to be sure that they are the ones to invent new products and processes to insure the survival and prosperity of the organization. Maintenance is no different. Maintenance needs an ongoing investment in research and development to improve the delivery of product.

What is at stake could be the survival of your organization (if not of outsourcing your department, or at a minimum, losing your job). There are competitors of all types from all over the world (no off-world competition –yet) that are eyeing your market share and they are not standing still.

Many sections of this book deal with increasing the productive output of your plant. In this section we will give a framework for all improvement projects. There are opportunities in every maintenance operation. The most interesting fact is that the people in the best position to know where there is waste in a factory are the machine operators and the maintenance workers. The waste literally (in some examples) drips through their fingers.

An internal study done by a major maintenance provider in Canada estimates the opportunity:

Percentage of possible savings of maintenance budget dollars

39%	Re-engineering and maintenance improvements
26%	PM improvement and correct application of PM
27%	More extensive application of predictive maintenance
7%	Improvements in the storeroom

First let's agree on what continuous improvement means.

Continuous Improvement means either ongoing reduction in:

1. Labor (operator, mechanic, and contractor)
2. Management effort (reduce headaches, and non-standard conditions requiring management inputs)
3. Maintenance parts, materials, and supplies
4. Raw materials (reductions in scrap for example)

5. Energy (electricity, gas, etc.)
6. Machine time (faster cycle times, fewer breakdowns, reduced set-up time)
7. Capital (less money needed by having, for example, fewer machines)
8. Overhead (less staff, smaller factory)

-or-

1. Improved output, reliability (uptime)
2. Improved repeatability of process (quality)
3. Improved safety for the employees, the public, and the environment

Continuous improvement has five essential elements

The five elements are Commitment, Measurement, Information gathering, Investigation, and then Action. In the sections that follow we will look into each area.

Commitment: In traditional maintenance department settings, the commitment to improve is personal. A person (rather than a department or plant) sees a way to improve maintenance or operations. Although this exercise is great and satisfying for the person concerned, it will not have enough overall effect to make a difference. To have enough effect we must make continuous improvement into one of the core activities of everyone's day (giving us the third curve if the improvement is directed toward reliability improvement).

Organizations that are committed to continuous improvement commit the time to do the analysis necessary (in preparation work – like in painting a tank, most of the work is preparing to paint and cleaning up afterwards). These organizations have established that appropriate effort in continuous improvement provides a substantial return on investment. They allow and encourage maintenance workers to participate in problem solving efforts that involve other departments of the organization including production, engineering, quality, cost accounting, and marketing. Effective commitment also is a long term choice. In good and (especially) bad times the maintenance department must always be looking at improvements.

Measurement: How do you know if your great idea actually saved money and where did the money go that it saved? A prerequisite to continuous improvement is establishing ways of measuring the inputs into the process (such as the maintenance effort), as well as measuring specific areas where improvements are taking place (like energy savings or scrap reduction). Without measurement, it is difficult to determine if an operation is truly improving.

You may be doing great in your marketplace due to factors that are outside your control. For example, when the currency in your country is weak your

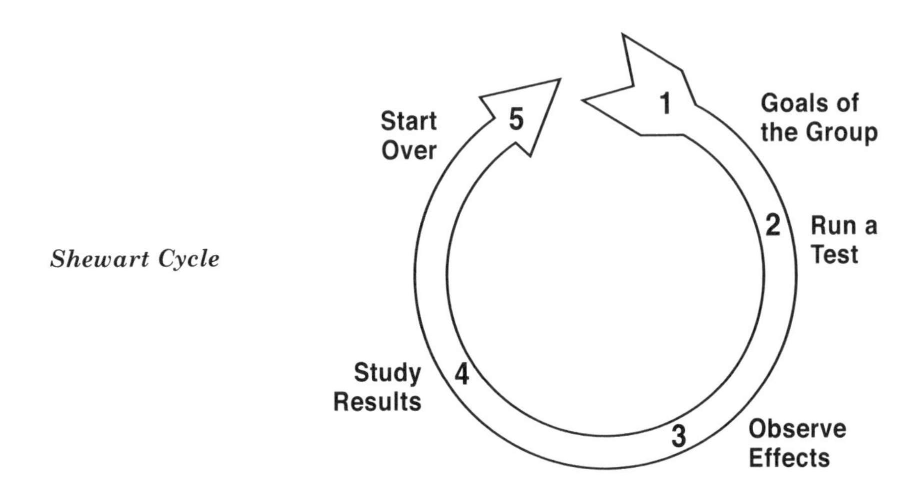

exports may go up because your products are cheaper in the currency of your buyer's country. That is the time to grab market share.

If you want to change the competitive landscape of your market it is essential to use the high volume times to invest in lowering your cost of production. Be careful because if you don't, the first "correction," in exchange rates will wipe out your market share advantage. Your global competitors might have made real improvements to stay in business (to compete with your currency advantage) and now, with the currency advantage gone, you could be wiped out.

The process of setting up measures is called Benchmarking. For the big areas (such as maintenance department efficiency), there are three types of benchmarks used by maintenance departments. For smaller improvements in processes or tactics, only the first type of benchmark in used.

1. Internal or historic benchmarks are based on prior periods and are the most common and by far the easiest to set up. An internal benchmark might be downtime hours due to maintenance problems. This benchmark could be tracked monthly (or even weekly) over a long period of time. Continuous improvement could then easily be measured against prior periods. All measurements have this internal method somewhere in the report.

For improvement projects, the internal measurement might be started just before the experiment to establish a baseline. The measurements might be repeated after the improvement to establish that improvement was indeed achieved.

2. Best-in-class is a benchmark of the best in your industry. Some large organizations will take the best plant of a certain type (if they have many similar plants). Other organizations will review trade literature or run a study to determine the best plant in the business, then compare themselves with that plant. A benchmark might be the number and severity of customer complaints per month. Compare yourself with the best plant of your type in the industry. Continuous improvement would measure your progress in catching up with and eventually surpassing them.

Although this process is expensive and sobering, it is very useful. In one instance, a plant had an OEE of less than 45% for a particular type of equipment. They hired a 'hot shot' that worked with them and improved the OEE to just under 60%. The plant manager and other staff members were very impressed and satisfied with the progress. The satisfaction quickly left when they found out that a similar plant under similar conditions (non-competitive geographically) routinely ran 80% OEE. That superiority means that for every dollar of assets (machinery) the better company produced 25% more product. That kind of advantage can outweigh some of the advantages of lower labor rates, high value currency, or whatever your barrier is now.

Best-in-class is a great motivator to pursue improvement projects. The results can also show where the opportunities are by demonstrating that improvements are possible.

3. Best-in -the-world is the ultimate comparison between functions. A best-in-the-world benchmark might evaluate your telephone answering function and compare it against the best in the world (such as Federal Express or Lands End retail catalog sales). The comparison organization might well be in a vastly different type of business (or even in government or education). Continuous improvement puts your organization alongside the best there is. Your achievements against the best in the world's benchmark are tracked and reported upon. The advantage of the best in the world comparison is that those concerned may never have heard that something cannot be done.

Information: Information is the core of the process. To achieve your improvement goals you will have to examine all production data, all minor jam-ups, all failures, all short repairs, all PM activity, and all other maintenance events for opportunities to reduce the inputs or help the improvements:

Conduct an Investigation: If we were looking at a series of costly breakdowns we would review the production numbers and then the maintenance incident history (an incident could be a breakdown, a series of breakdowns,

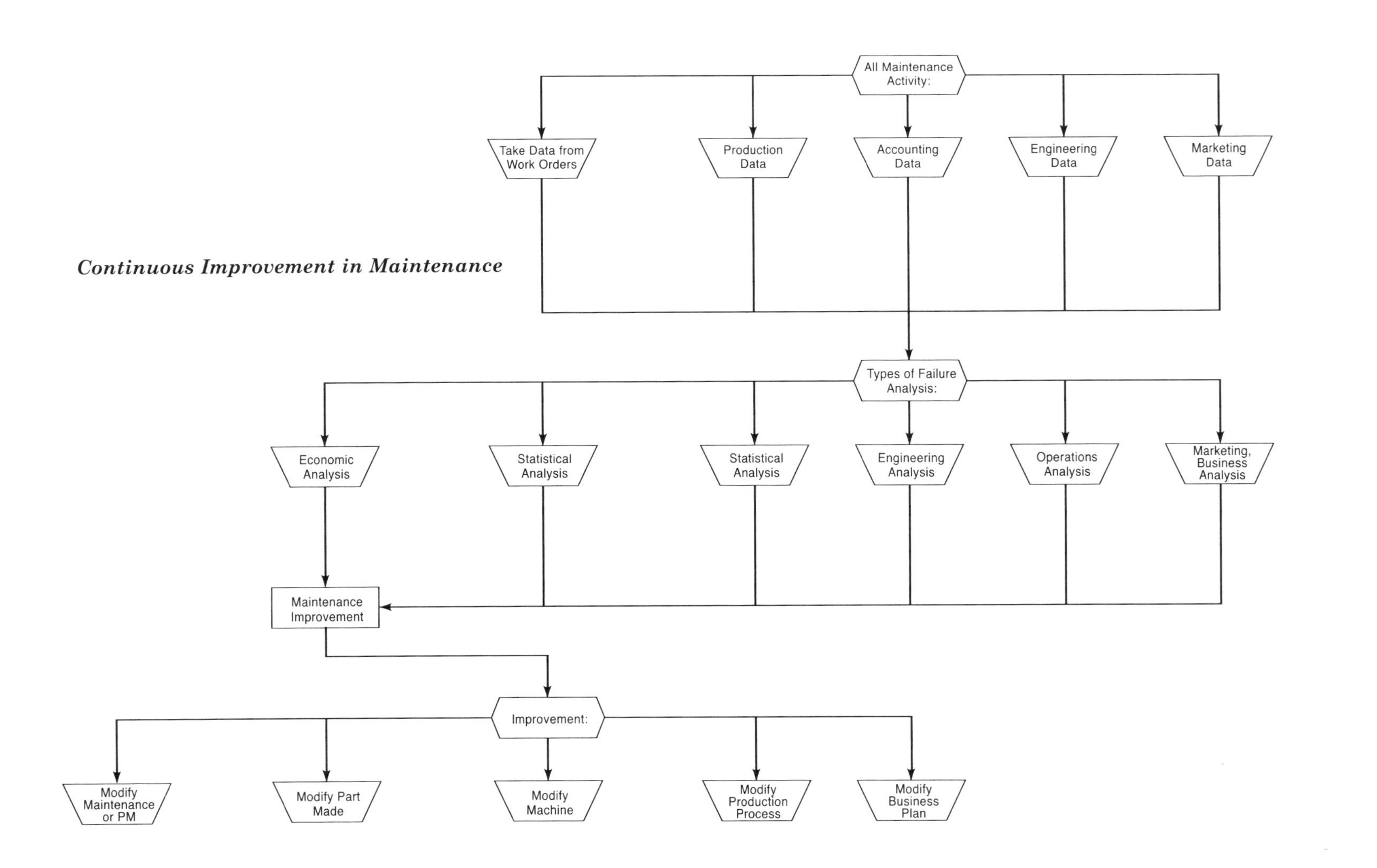

Continuous Improvement in Maintenance
All Maintenance Activity:
Take Data from Work Orders
Production Data
Accounting Data
Engineering Data
Marketing Data
Types of Failure Analysis:
Economic Analysis
Statistical Analysis
Statistical Analysis
Engineering Analysis
Operations Analysis
Marketing, Business Analysis
Maintenance Improvement
Improvement:
Modify Maintenance or PM
Modify Part Made
Modify Machine
Modify Production Process
Modify Business Plan

PM's for a machine, a series of minor adjustments, or other maintenance activity). The asset, area, or system needs to be reviewed from six different points of view:

1. Economic analysis

A. What is the cost of the incidents?
B. What is the cost of the downtime?
C. What is the cost per year?
D. What cash flow is generated by an improvement project (that is available to fund the improvement)? Cash flow numbers will answer the important question of how much should we spend to fix this?
E. What is our investment in this asset or process?

2. Maintenance analysis

A. How disruptive is this breakdown (to other jobs, to our people's home life, etc.)?
B. Does this breakdown cause mechanical or electrical problems elsewhere?
C. What is the honest opinion of the maintenance old timer' experts?
D. Is the root cause a faulty or inadequate maintenance procedure?
E. Is the root cause related to inadequate training in any maintenance skill?
F. Was there a failure of the PM system?
G. Do the PM task lists look at this failure mode?

3. Statistical analysis

A. How often does the incident occur?
B. Is there a pattern or trend?
C. Is the pattern or trend deteriorating (or improving)?
D. What is the mean- time-between-failure (called MTBF)?
E. What is the mean-time-to-repair (MTTR)?
F. Can the MTBF be managed by replacing the component more frequently than the failure frequency (using PCR)?

4. Engineering analysis

A. What (really) happened? (If it can ever be known.)
B. Is a structural analysis of the broken parts indicated?
C. What happened just before the failure?
D. Why did the breakdown take place?
E. Are we looking at the root cause or a symptom?
F. Is the root cause related to an inadequate spare part specification?

5. Operations analysis

A. How do these events impact operations?

B. Does the event force the failure of other parts of the process?
C. Can we bypass the problem with a back-up or standby unit?
D. Is there a scrap, rework or start-up exposure?
E. Do we have to dump intermediate product?
F. Is this event indicative of a failure of the operations system?
G. Is inadequate operator training a cause or a contributor?

6. Marketing/business analysis

A. What is the impact of this failure on the customer (internal/ external)?
B. Can we afford to have this type of event happen?
C. Can this event impact quality?
D. Does this event impact morale?
E. Is there an impact outside our sphere of influence (environmental, competition)?
F. How high should the priority be to deal with this problem?
G. Is there a regulatory or legal impact to this event?
H Is there a logical business decision indicated (outsourcing the part, sell off product line)?

You can see from the scope of the list above, that no matter how knowledgeable individual maintenance professionals are, they cannot know all the ramifications of a major series of maintenance events. If continuous improvement is pursued seriously, multi-departmental teams will be necessary on an ad hoc basis to attack problems.

Action: Based on the investigation, institute improvements in one or more of 5 areas. The trick is to have a complete investigation. In many instances, the solution will be clear just from the facts of the investigation. In many other examples, the action taken is designed to gather information, test a theory, or even (when you're getting desperate), play a hunch about the nature of the problem.

1. Modify maintenance practice or PM procedure

Add tasks to catch the particular failure mode earlier on the critical wear curve. Increase the technology of the tasks such as adding a Shock Pulse Meter or vibration analysis. Increase frequency of the tasks. If economic and business analysis shows that we are spending too much on PM in relationship to other costs do the opposite of the above (reduce frequency, depth, etc.). Investigate better cleaning, improved lubes, better alignment processes, if this is a wear problem. Design easier to do maintenance procedures and re-engineer the equipment to suit.

2. Modify machine (maintenance improvement)

Improve the machine so that it doesn't break or need adjustment. Improve the tooling. Make it easier and faster to do the maintenance tasks. Automate some of the maintenance tasks. Remove the source of the problem (redirect the dirt so it doesn't fall on the cylinder). Reduce (or increase) the number of steps needed to make the product. Add automated lubrication systems. Add instrumentation to the machine. Increase the size or quality of bearings, seals, or wear surfaces. In short, reengineer the machine so that it needs less attention.

3. Modify part/product

Make the part easier to produce. Improve the tooling. Change the shape, size, material, or finish of the product. Reduce (or increase) the product's specifications.

4. Modify production process

Improve the whole process. Improve incoming materials. Improve the process to allow greater variation in incoming materials. Make the transfers between processes more bullet proof. Increase yield. Look for improved technology. Add robots or other automation to the process. Look for a whole new and more reliable process. Give headache processes to vendors who are expert in that process.

5. Modify the business or marketing plan

Discard the product, increase the lead time, buy your competitor, sell out to your competitor, raise prices, sell custom units, have your competitor make your parts/you make theirs, hire a top engineer/production person or consultant to help smooth the process. Outsource specific operations or the whole process (stick to what you do best).

There is one big 'but' in the world of actions. Actions have consequences. Actions might also have unintended consequences that are far worse than the original problem! Look closely at the action and insure that there will be few, tolerable, unintended consequences.

Goals

Both discontinuous improvement (for example the invention of a disruptive technology) and continuous improvement are the only antidotes to the constant pressure of competition. Being steeped in the concrete reality of the equipment, maintenance would naturally be the center of the continuous improvement program.

It is essential that management realize that continuous improvement in the maintenance department is everyone's business and can only be achieved with everyone's input. The goal of continuous improvement is the gradual elimination of the need for maintenance. The inputs drop and drop until maintenance is non-existent. Although I don't expect this outcome soon, the goal is there to work toward.

When continuous improvement is part of the everyday work small investments can result in major savings. In one maintenance department with 135 maintenance workers a $75,000 investment in training about continuous improvement resulted in $750,000 of savings in the first year. Even then, only a small number of projects were tried. Here is a sampling of the projects, whose simplicity is striking. .

Eight Examples of completed projects

Gold * Positive effect on A/C electric consumption of increasing the set point of an office from 72° to 75° F. No complaints were registered. The electricity usage dropped by 10% resulting in a $7000 year savings per office with an immediate total approaching $315,000 for the office complexes. It was given a Gold star because there is little investment, large potential savings, and immediate returns.

Silver* Impact of the use of stabilizer on the consumption of chlorine. The team added $225 of stabilizer to a reactor. They charted chlorine usage for 1 week before and 1 week after stabilizer was added. Chlorine usage during the test period dropped by $80 per week. Potential savings is $4000 per year per unit. Silver Star is awarded because investment is small but the payback is large and immediate. This technique can easily be applied to 75 reactors where stabilizer is not being used.

Bronze * Impact of Ever-pure filter system to prevent corrosion failure in small boilers. The manufacturer increases the warrantee from 1 to 7 years if an Ever-pure filter is installed. Installation parts and labor are about $550. Current failure rates of 1 liner every 2 years will result in a savings of about $2684 in 7 years for each unit. 14 proposed Viking units would yield $37,500 in avoided maintenance costs over the 7-year warrantee. In addition there are over 50 boilers of other sizes, makes and models with equally high failure rates. The filter might help generate an additional savings of $134,200. This project was chosen because of the ease of the savings (from a warrantee), the real reduction in labor, and the high cost of parts. If the technology proves to be effective, the project can be expanded to all the boilers in the plant.

Compare circulating pumps. Project compares an existing cast iron circulating pump to a smaller and lighter stainless steel pump. The smaller pump has been in use in the plant for 7 years. Analysis of replacement cost, reliability, energy usage, and complexity of installation shows that the smaller stainless pump is clearly superior. It costs $1500 less to purchase, is more reliable, and uses about $70/yr less electricity. Recommendation: replace all circulating pumps with the stainless one as they fail.

Impact of relamping and cleaning fixtures on light output and electric usage. In an office where there were lighting complaints, the team replaced the tubes (with efficient ones), and ballasts (with electronic ones), and cleaned the fixtures. The candlepower at desk level increased from 12.5 to 56.8 (standard is 60). The amperage consumption at the breaker went from 14 to 8. Total cost was $305 including direct labor at $25/hr. Electric savings were $220/year. Several ballasts were leaking and were hazardous. Recommendation: create an annual campaign where each floor chooses its worst 2-3 rooms for lighting. Teams will then re-lamp, re-ballast, and clean fixtures. Based on the other relamping project high efficiency ballasts and utility rebates should be looked into.

Testing premium cutting blades verses currently used inexpensive blades. Premium blades cost $8 each and the current blades cost $2 each. The blades were tested under real conditions during alternate weeks. The conclusion was to stay with the current blades used by the department.

Comparison of individual pothole repair with parking lot resurfacing. Calculations show that resurfacing is economic if there are more than one pothole per 1000 sq ft. Pothole repairs are more costly, they last only a short time, and are usually done on overtime (to minimize interference with the operation). Resurfacing lasts an average of 6 years and enhances the look of the plant.

Comparison of direct drive exhaust fans with belt driven fans. Study shows that belt driven fans are more susceptible to maintenance calls and noisier. Recommendation: In new installations, direct drive fans would be less expensive to purchase, quieter to use, and require lower maintenance input over the life of the unit. Replacing existing units is not indicated unless extensive other work is needed.

World Class Maintenance Management

The wave of organizational changes required to run a comprehensive quality program has hit the maintenance department. Consistent quality depends directly on the maintenance department. Maintenance professionals the world over have heroically struggled against inadequate support, old equipment, inadequate engineering, little time to test processes, excessive run time between services, high tolerances requirements, and uninterruptible utilities.

This wave of change is great news for maintenance. Maintenance wisdom has long been restricted (to maintenance professionals), and ignored. Maybe now, the maintenance people will get the respect they deserve, provide unique insight, and receive the support they need.

Competitiveness in world markets and survival now drives the changes in how top management views maintenance. A World Class Maintenance department enhances their organization's ability to provide their product or service. The maintenance department becomes a strategic asset of the organization rather then a necessary evil.

The twenty attributes listed below are derived from the author's visits to many maintenance facilities and the excellent work of W. E. Deming, Don Nyman and others.

1. Top management has awareness and appreciation of the significance of maintenance to the overall objectives of the organization.

The first key to world-class maintenance effort is the support, understanding, and trust of top management. These factors are frequently missing. One mission of the world-class maintenance department is education of top management. Although much of maintenance is unknowable (the exact minute a gear will crack), much of maintenance reality (machines do break down if they cannot be serviced) can be taught. Only a shortsighted executive would ignore the dire warnings of the finance Vice President. Yet many Managers routinely ignore the warnings, knowledge, and experience of their Maintenance Managers.

As an example, a power utility company builds a generating station in the Midwest with the design specification that it will be able to have 85-95% availability. To support this level of availability the designers have long and detailed meetings with the maintenance leadership. Every failure mode is discussed, looked at, planned for, and (where possible) designed out. The capital spares inventory was 24 million dollars (just reduced by the maintenance department to 15 million). This utility realizes that profit in power generation hinges on the wisdom of the maintenance input to the design and operation of the plant. This is an example of support, appreciation, and understanding of the role of maintenance in the success of the operation.

2. Mission statement

A mission statement is necessary to run a world-class maintenance department. The mission statement becomes the primary reference point that the staff can use for effective decision-making. The mission statement covers the issues of customer orientation, continuous improvement, quality, safety, environmental position, employee development, down time, proactive stance, and any industry specific areas.

In the executive conference room of a vehicle manufacturer was a beautiful poster titled Mission Statement of the Corporation. This poster cost in excess of $5000 to produce and print. I asked a class of maintenance supervisors sitting in the room at the time "did the company have a written mission statement?" At least half said no and the others thought they did. No one in the room could remember any of the statements of the document. This is an example of a fake mission statement since it will not guide anyone toward better quality, safety, or anything.

The mission statement defines the organization in concrete terms. This definition is particularly useful for a trade's people in the middle of the night, far from the other workers, or for some one new to the company when there is no supervision to ask. The mission answers such questions as how much effort do I put into quality, safety, etc.

In organizations that are serious about their mission, everyone knows the critical items on the mission statement. If you doubt this, ask a Dupont employee about safety, or a Saturn employee about product quality.

3. Constancy of purpose (emphasis on long-term goals and views)

There is nothing more debilitating to the effectiveness of a maintenance department than the management `flavor of the month club'. That is, a management that jumps on the bandwagon of every new guru and loses the focus necessary for world-class performance. Last year it was empowerment, this year it's stay close to the customer, and next year it will be we worship at the altar of quality.

The march of equipment and facility deterioration moves at a slow, inexorable pace. The march of deterioration can be slowed by good lubrication, cleaning, and bolting practices. The effects of the march can be treated by effective corrective maintenance. The results of this march of decay can be detected by predictive maintenance and delayed by preventive techniques. But no technique can stop the march entirely.

Upgrading a reactive department to preventive, then to predictive, then to World Class is multiyear effort. Great achievements like world-class maintenance require a long term steady hand.

This attitude should be reflected in a long-term budget. A five- or even ten-year budget is needed for maintenance and rehabilitation, and will provide information to the corporation as to future needs. Many of the assets that maintenance supports have long decay cycles. These long cycles respond best to a constant approach where we look, record; and document the deterioration for years while planning an intervention. The intervention is then well conceived, reasoned, planned, investigated, and executed.

4. Both maintenance and management have patience

Going hand-in-hand with constancy of purpose is patience. Maintenance problems take a long time to develop and consequently they may take a long time to fix. Substantial maintenance system and procedure improvements require time and investment. Initial results often take a year or more. Managers used to dealing with monthly results will be uncomfortable with the longer time spans.

Steven Covey, a leading business thinker, speaker, and consultant, sums up the paradox in saying that there must be an unchangeable core at the center of the organization to be able to react quickly and effectively to the marketplace. This core covers the essential values that make the organization successful in the marketplace. The marketplace may make different demands month-to-month but the core values stay the same.

One of the core values is the way we treat the people and assets under our stewardship. That value is defined by the organization's attitude toward maintenance and maintenance leadership. Patience is also an expression of trust.

5. Focus on service to the customer

Another core value of the organization is excellent service to all the customers. The maintenance department is a service vendor for all other departments of the organization. These departments are the customers of the maintenance department and need to be understood through regular communication, periodic surveys, and needs assessments. When maintenance efforts fail it is the customer that suffers. Every member of the maintenance work group should be familiar with the impact of their actions on the customer.

No maintenance department in the over three thousand that have attended the author's classes has ever gotten bad feedback from attending to the customer too well. Maintenance brainpower would be well spent if it concentrated on faster delivery of higher levels of service. The maintenance department must strive to serve the real needs of the customer.

Some effort would be rewarded for determining what maintenance changes could impact the outside customer (called the big C to distinguish it from the internal customer called the little 'c'). How can what you do, impact them and their business? How can you make your outside customer's life easier? It might be hard to see what you can do to help the outside customer beyond doing your job inside but the rewards for success are great. Any substantive improvements will enhance the competitive position of your whole organization.

Many maintenance organizations take a superior attitude toward production (the customer). They say, "We are the real brains and production is a bunch of idiots". They can prove it by showing you all the stupid things that happen in production. These departments miss the fundamental point which is that our purpose flows from service to the customer. Maintenance leadership sometimes misses opportunities to inform maintenance workers that their attitude affects the service they provide.

6. Proactive not reactive

The world-class maintenance department does not allow critical assets to deteriorate to the point of a breakdown. The proactive maintenance department does not wait for the breakdown but goes out onto the plant floor looking for impending problems. Implicit in the proactive approach is the will to take equipment out of service for repair before the breakdown occurs. The reactive approach says `if it ain't broke, don't fix it.' The proactive approach says `don't let it break down period! '

Examples of proactivity include inspection, cleaning, tightening, lubrication (all of which are PM activities), complete testing of new equipment, operator certification programs, continuous training programs, well thought out store rooms, reviewing designs before construction for maintainability, etc. In fact, all activity related to avoiding breakdown in the future.

Consider the attitude in a program to preserve assets in a museum of art. Of course, an art museum has a great restoration department to repair damage after it has been done, but the bulk of their activity is in creating an atmosphere that will prevent deterioration in the first place. Maintenance departments need to get in front of the action out on the shop floor and not wait for bad things to happen.

There is another aspect to proactivity. A proactive maintenance department will spend a significant percentage of its assets on maintenance prevention. The most proactive stance of all is to eliminate the need for maintenance.

The toughest aspect of proactivity is self-discipline. You must maintain your proactive activity without feedback and support. In fact, proactivity requires doing things that may seem irrational to outsiders, without a long explanation.

7. Root cause analysis

Getting to the root cause of a problem and fixing it is the best way for a maintenance department to gradually improve the delivery of maintenance service to the customer. Root cause treatment comes in two stages.

Most good mechanics will work on a breakdown until they understand and repair the root cause. This first stage distinguishes the real mechanic from the parts changer. Root cause analysis requires some time to be spent on study of the system. Many organizations imagine that they do not have the time for study, and they force premature decisions on the mechanic. I say `imagine' because they often seem to have the time to repair the same problem over and over again (rather than taking the extra time to understand and make a more permanent repair).

The second stage is to re-engineer the system to avoid that mode of breakdown in the future, which may be well beyond most mechanics working on complex systems. In large factories the competition, the severe service requirements, and the sheer cost of maintenance require higher levels of expertise.

Root cause analysis can be done by anyone who has knowledge about the prevailing situation and ideas about elimination of the problem. Root cause investigations are an excellent way to use the team concept and break down interdepartmental barriers (building teams with maintenance mechanics, engineers, customers, etc.).

Root Cause Analysis is a merger of reactive and proactive skill sets. It is generally initiated by a failure, making it reactive. It requires attention to equipment history, operating conditions, engineering, and the details of the failure, which are proactive activities. The mechanic who successfully undertakes this analysis has the ultimate cross training experience (and has great fun besides).

8. Team concept

Many maintenance problems today are too complex to be solved by a single person. World-class maintenance departments recognize and capitalize on the different skills and expertise of different members of its crews and the other departments. Teams are used extensively to solve problems, plan jobs, and institute improvements. A single maintenance worker might be involved in several teams simultaneously. Some teams might be ad hoc (set up for one problem) and others might be standing teams (for safety, environmental, etc.). Leadership training and opportunities safely to practice the exercise of leadership are part of the team concept.

One of the most interesting examples of team projects was in the automotive industry. Automobile companies routinely buy competitor's products and disassemble them for study. In one company they pull together a volunteer, ad hoc (one time, as required) team consisting of design engineers, people from general assembly, maintenance people, and even administrators. This team would take the car apart and attach the parts to boards for display. They would analyze the part counts and assembly techniques, and prepare a report and presentation for the larger, new car design team. The assignment is a desirable one because, in the words of one maintenance electrician, it is fun.

9. Fading of traditional interdepartmental barriers

Hidden away in many other departments in the organization is a significant amount of the expertise needed for successful maintenance. The world-class department taps into these storehouses of expertise by breaking down interdepartmental barriers. Traditional departments that support maintenance (engineering, stores, safety, purchasing, and housekeeping) are actively involved in maintenance issues.

Non-traditional departments (finance, cost accounting, data processing, marketing, strategic planning) are also brought into the Maintenance decision processes. The team idea includes interdepartmental teams with significant input from the different departments. For world class maintenance to take hold and flourish, detailed maintenance knowledge must cross departmental boundaries.

Part of the goal of destroying these barriers is the distribution of maintenance knowledge throughout the whole organization. Unfortunately, under old departmental structures, maintenance knowledge is not well distributed throughout the organization.

The result is ill- advised decisions such as that used for an oil refinery mothball project. When oil prices fell, an oil company decided to mothball one of their refineries. The accounting department estimated that the savings would be $10 to $15 million per year. They further estimated that the refinery would cost $75 to $100 million to put back on line. To maximize the savings they laid-off the entire maintenance staff and just had security personnel on the site. After 8 years the price of oil recovered to the point that they wanted the refinery back on line for some large contracts.

Because of a lack of basic maintenance knowledge and of effective communication, the refinery was almost a complete loss and cost almost $700 million to bring back into service. Maintenance knowledge would have alerted accounting that a small investment of $2 to $3 million per year would have preserved the plant. Eight years without investment caused significant deterioration. Of course, the company would have had to spend $16 to $24 million over those years. But the return on investment for that money would have been well worth it.

The rest of the effort in reducing barriers to wider use of maintenance must be directed toward introducing sophisticated cost accounting, finance, engineering, and other skills into the maintenance department.

10. Customer participation in Maintenance (with training!)

World Class Maintenance requires the operators' to be involved. The more the operator is involved the better for all concerned. The benefit for the operator is the feeling of being the owner of the process/machine, improved responsibility, and higher total productivity. The operator is the logical person to perform certain basic PM tasks because he/she is in daily contact with the machine. The machine is really theirs.

The benefit for the company is improved tracking of responsibility, improved quality, and improved knowledge, which leads to improved productivity. A knowledgeable operator who feels responsible will make better parts at a higher rate. Downtime will be reduced and small problems will be addressed quickly.

If we did all the PM's we should do, we would soon find there is more maintenance to do than there are maintenance people to do it. We need an additional resource. The largest hidden resources of the maintenance department to service the assets are the users themselves. In some industries (such as the trucker checking his/her own oil and doing the pre-trip inspection) it is commonplace for operators to participate in some level of Preventive Maintenance procedures. In fact, the pre-trip inspection is the law in trucking and aviation. In other industries, operators don't touch the equipment except to operate it. The operator just pushes the button or watches a gauge.

This ownership takes another step in some departments where the operator acts as a helper on large repairs or as a safety watch person if confined space entry is required. Once properly trained, the operator can be a great asset. The best implementation of this idea in factories is called TPM (Total Productive Maintenance).

11. Cross training (also known as multi-skilling)

High levels of productivity require some level of cross training. Cross training simplifies job planning. Less time is lost coordinating work done by different crafts. Many maintenance departments have enough people but have inadequate staffing in particular crafts or skills. Cross training also provides better job-security because of the possibility of changing to jobs where the craft is more in demand when times are slack.

The major reasons for cross training are to improve productivity and allow one person to do more of the job so that they can feel ownership. A powerful motivator of the maintenance worker is the feeling of pride in a job well done. A cross-trained worker is more likely to feel the pride because they are doing the whole job.

Multi-skilling depends heavily on a successful training and testing program. In all multi-craft shops, training is an initial issue. Many firms bring in outsiders, tech schools, or other training professionals, to organize the massive training effort. Some of the best maintenance departments pay craftspeople for qualifications in extra crafts.

In an aluminum mill in northern Alabama there are 11 crafts in the maintenance shop. In a worst case scenario it could take four people and their supervisors to change a small motor. We would call out the pipe fitters (if we needed to disassemble or change the piping), sheet metal mechanics (for modifications to the shroud or cover), an electrician to remove the wires, and the trusty millwright to remove and replace the motor. The quality of each craft would be great, but the productivity would be well below world class standards.

12. Continual training

Your factory has increasingly sophisticated technology. The technology is 1980's through 2000's vintages. Your crew was last in formal training 15, 20, 25 years ago. This gap must be filled by updating skills through continual training. The alternative is lower and lower quality, increased downtime, and a growing inability to perform even routine root failure analysis.

You and your staff are involved in continual training. This training is an investment that organizations make in their major assets (people). Training has four steps (if followed they give the most results from the least training dollars): First, analyze the job for needed knowledge, skills and attitudes. The second step is to evaluate the candidate's need for training against the job's requirements. The third step is to develop a training prescription. Finally, when the training is completed, the candidate has to be post tested to see if the material necessary for success in the new job was learned satisfactorily. Traditional `shotgun' approaches are too expensive and not effective. The training process is ongoing because the changes in technology and gains in products and procedures do not stop.

In a new plant for automobile production, the standard for training is 96 hours per year for all people. A high tech manufacturer in the upper Midwest requires 5% (104 hours annually) of all direct hours be spent in class of some type. These firms realize that such investments pay interest in the ability of the craftspeople to adapt to new technology, new processes, and new organizational structures. Topics could include engineering, craft skills, multi-craft skills, computerization, maintenance management, safety, or your industrial process.

13. Information sharing

Information essential for effective maintenance exists in many locations in any organization. Maintenance touches many different levels of activity and each level has important information. Some examples of critical information that impact maintenance decision making are fixed asset accounting methods/decisions, overhead costs, downtime costs, equipment retirement cycles/budgets, interdepartmental priorities, and more. World-class maintenance can happen only in an atmosphere of open exchange of financial and production data.

Without critical information the maintenance department is out of the loop and cannot make effective decisions. Worse than being out of the loop, maintenance can make good maintenance decisions that still work against the organization.

For example, in a Maryland based plastics extruder, a maintenance worker playing around with a new vibration analysis tool found a problem with the gear used to drive the main extruder screw. An immediate repair to prevent breakdown would cost $500. After breakdown the bill would have jumped to $5000. He was very proud because the savings was estimated at $4500.

He discussed the plan with the shift manager and the maintenance manager and they decided to do the repair immediately on 2nd shift and go for the savings. The next day the manager walked into the Wednesday morning production/ maintenance scheduling meeting. He was greeted by a chilly silence before the meeting. Production had missed a just-in-time delivery for a new customer that morning. They asked "did he know why unit 5 was taken off line for an entire shift and a half?" Almost $40,000,000 of new business was put at risk. Lack of information (lack of communications) created a costly blunder.

14. Benchmarking

The benchmark is the mark that old world craftspeople made in their benches to use for measurements. All parts they made were compared with the benchmark to assure that they would fit. Today a benchmark is the standard for performance and means of measurement. We use benchmarks to tell if an operation is improving, stagnant, or declining. A world-class department wants to know how it is doing.

One of the problems of maintenance has been the difficulty of finding benchmarks that could be used effectively to measure the performance of a maintenance department. Some measures that can be effective would be the number of completed work orders versus the number of incoming WOs, PM hours to total maintenance hours, percentage and quantity of emergency work, number of callbacks, downtime (or uptime including downtime reason), production quantity/quality, maintenance cost per product shipped, and many others.

15. Continuous improvement

Many factors contribute to an organization's survival. Continuous improvement is one of the big contributors. Continuous improvement is everyone's job. The world-class challenge is to produce the same or greater output at a higher quality with fewer inputs. One of the inputs is maintenance effort and parts. You should establish the inputs needed per unit of output (output can be measured as cars assembled, cases of soda bottled, barrels of oil refined, etc.). Through your ongoing continuous improvement processes the input per unit output should drop. Last year's numbers can be the benchmarks to beat for this year.

Continuous improvement is also an attitude. Many maintenance people fall into a dangerous rut where, when a system is understood and its failure modes become familiar, they feel satisfied. They feel as if they can handle anything that might happen. They are not at risk. While it is nice to feel as if you can handle anything that might happen, it is deadly to accept the status quo. You have to fight this tendency to maintain the attitude of continuous improvement.

16. Attachment to the people rather than to technology or computer systems.

We run the risk of thinking that if only we had a new computer, bar coding with a new scanner, or other tool, our maintenance department would finally get better. The truth is that most good maintenance practices are basic and low tech. Having our everyday people, our first line maintenance workers, do the basics well is critical to management of breakdowns. Our people are the important asset. Our thoughts toward improvement should always look to their needs first. Any system changeover should consider good employee treatment, adequate adjustment periods, and sufficient training.

We routinely see systems that have run amok. Such systems would spit out PM tickets for equipment retired a decade ago. There seemed to be no one left that could change (or wanted to change) the file to get rid of the irrelevant PM. To add insult, the PM tickets in question have to be closed out as complete or all the reporting will be erroneous. In another example, the maintenance manager got a 1400 page weekly maintenance report. It included everything for all divisions mixed together. The data processing function could not or would not suppress the extra information.

17. Attachment to people so that every other option is looked at before layoffs (W.E. Deming says drive out fear).

Many organizations lay off people as a first resort rather than a last resort. Mindless expansion and contraction of the permanent work force is immoral in today's competitive world. Every other option should be tried before layoffs including shortened workweeks, slashing executive salaries, using maintenance for construction work, accelerated retirement, transfer of maintenance to production, staff, even marketing jobs. Companies need new strategies to cope with expansion, contraction, plant closings, and changes in the product mix.

It is quickly becoming clear that the true asset of a manufacturer is the knowhow of the employees. This specific knowledge is essential in all aspects of modern management. Processes cannot be improved, products cannot be made more consistently, and benchmarks cannot be achieved without specific knowledge. This knowledge comes from years of solving problems in a factory or an industry.

In all theories of motivation, when a person's paycheck is threatened, they are not available to perform their highest quality work. It is difficult, in an anxiety-producing environment, to concentrate on the details that create quality and safety.

18. Willingness to run controlled experiments

Controlled experimentation is the key to new knowledge. Using the Shewart cycle or other technique, ideas are introduced by the work force, tested, refined, retested. Root cause analysis (what is the root cause, how can we fix it) will also suggest needs for experimentation. Controlled experiments not only support continuous improvement but they are the key to finding effective improvement ideas.

Every maintenance budget should have some money set aside for maintenance experiments. A good starting figure might be 1%-2% of the regular budget. Returns on investment should be tracked and successes publicized.

The only way to improve is to try different ideas, technologies, techniques, approaches, etc. Coupling experimentation with job enlargement and training will propel your maintenance department to the forefront of its field.

19 Application of statistical tools to maintenance

In the great move toward quality in the late 70's and 80's production and top management people discovered some old statistical tools. These tools would explain the problem of natural variation. They also would help identify when a process went out of control. Maintenance can learn from statistical thinking with use of failure analysis, PM intervals, replacement life, and other approaches.

The excellent reliability record of the commercial aviation industry is due, in great part, to the application of statistical tools to aircraft reliability. Much credit for reliability goes to the aircraft companies, component suppliers, and engine manufacturers. The actions that really keep the planes flying, even after 20 or 25 years, are the refurbishment cycles, scheduled replacements, inspections, and the tremendous database of all incidents.

Statistical techniques are only as good as the database they are derived from and the reliability of the intervention. The aircraft database, for example, includes detailed plane configuration information, maintenance records, incident records, and utilization data on all the commercial aircraft in the US. Reliability assessment is based on billions of miles and hundreds of thousands of take-off and landing cycles.

Part of basic training for maintenance apprentices should be calculation of Mean Time Between Failures (MTBF), standard deviations, Performance-Failure curves, and how and when to use them. Our countries' PM systems would improve overnight if we could get rid of emotional PM intervals.

20. Self motivation

A self-motivated work force is the result of management doing hundreds of little things, and a few big ones, right. The principles of world-class maintenance will result in a self- motivated work force and an environment that is exciting to work in. Imagine having a maintenance department that outsiders want to work in?

There are great opportunities for organizations that put major effort into making products, understanding the real maintenance issues, and becoming experts in their activities. The marketplace will no longer tolerate amateur manufacturers. World class maintenance concepts will deliver professionalism as well as higher levels of performance.

Additional thoughts on world class maintenance

In the years since the first edition of this book was published, the twenty steps have morphed into six areas of interest, which include many of the twenty steps with revised contexts and definitions.

As always, there is bad news and good news. The good news is that there are few real rules in maintenance. The bad news is that there is no book that you can refer to that has the rules for PM, for work orders, or for anything else in maintenance. Some people spend their first few years fruitlessly trying to figure out the rules.

From time to time pundits (like me) have advocated rules or at least guidelines for world class maintenance. We trade on the idea that people want to know the rules of any game they choose to play. So, based on experience, based on (usually) a limited amount of research, we determine the "rules of the game." This arrangement is great; if you like an expert's rules, you can follow them, and, perhaps, derive significant benefit. But if the truth is told, any pundit's rules for world class maintenance are his or her opinions. They are opinions, not rules or laws, nothing more and nothing less.

So until the congress passes laws or the agencies publish regulations, the good news is that there are no rules. It should be known that in some industries there are rules, where Congress has passed laws or regulators have issued regulations. For the rest of us, if it works it must be right!

Six Areas of Interest (for the world-class aspiring organization)

Companies have developed a new attitude toward maintenance activity. The best of the best organizations realize that while some maintenance is inevitable, a proper attitude is necessary to minimize their exposure (in other words their downtime, parts costs, and labor costs). In short, they view maintenance activity in a special way, and you should also.

- Management wants a department that is more proactive and less reactive. Finding a problem, planning the solution, and solving, it is less expensive in the end then having the problem find you (over and over again).
- Don't wait for stuff to break and mess up production, get out there and find deterioration before failure. Use non-interruptive inspection techniques, or well-timed scheduled outage to detect performance degradation and potential failure points.
- Once something is broken, apply root cause analysis to fix the cause not just the symptom.
- Less repetitive maintenance effort is better. Problems must be resolved permanently

In the old days management people were cowards because they hid behind their ignorance

- These people wanted the results without the investment. So they didn't put their money where their mouth was. They read that PM was a good idea and asked for PM without providing, even, minimal support, or machine access. Management thought that if they yelled loud enough or set a big enough goal the machines would be motivated to not break down.
- What happened is that marketplaces became more competitive and left no room for amateurs. Judicious maintenance investment produced winners.
- Management today is listening to maintenance opinions and factoring those opinions into large business decisions.

- Like any change, if management people want improved maintenance they must fund it.

In the old days maintenance folks were cowards and hid what was really going on in the maintenance department from management. Now the spotlight is on maintenance and hard numbers are king.

- Why? Management wants to see what is really going on. Secondly, numbers are used to measure continuous improvement (or lack thereof). How interesting would the Super Bowl be if there wasn't any score? We run the maintenance Super Bowl but won't tell anyone what the score is.
- Productivity is now second to results. In fact, world-class organizations realize that high productivity is a function of doing many small things right. Management wants results (such as uptime, reliability). The world-class organization's management wants high productivity and more sophisticated management of maintenance from the maintenance leadership.
- Management is requiring analysis driven maintenance decision-making, not seat of the pants-driven maintenance decision making. The key is making rational decisions backed by data that can be reviewed (and understood) by a manager without a lifetime of maintenance experience.
- World-class organizations have an increasing willingness to use sophisticated tools of statistics, finance, and accounting, in maintenance analysis.

Great maintenance managers have realized that how people are used is the key to success

- Teams are used to solve problems and are dissolved when the problem is solved.
- Traditional departmental barriers (letting people see more of the big picture to make better decisions), are fading away.
- There is more customer participation in Maintenance (with training and with proper management)
- Information sharing today includes things like sharing charge back rates, machine part costs, and the maintenance budget (for starters).

In fact, almost all maintenance difficulties are really people problems masquerading as maintenance problems.

- If people are your main asset, then it is prudent to invest in your people through continual training to improve the size and effectiveness of the asset.
- It is vital to be damn sure that nothing within our control gets in the way of our people's concentration (fear of layoffs is handled perhaps with a smaller core crew and supplementation with contract help). Our commitment must be to people so that every other option is looked at before layoffs (W.E. Deming says drive out fear).
- Cross training (also known as multi-skilling) is a goal to improve both productivity and the personal sense of satisfaction
- Our attachment must be to people rather than to technology or computer systems.

What is the real mission of your department? This real mission might be at odds with the pretty mission statement over the door. In the best organizations there is alignment between the mission over the door and what really is said or not said in the small meeting and on the shop floor.

- Always: Focus on service to the customer or focus on adding value for the customer
- Always: Focus on safe operations for the maintenance workers, operations, the general public, and the environment.
- Always: Look for ways to reduce the costs of your operation.
- Always: Look for better ways to do business by cultivating a willingness to run controlled experiments.

Once again, the goal is the same. We are seeking a powerfully self-motivated workforce and excellent execution of maintenance activity.

Important Question: Is your company ready?

If your company is not ready, with increased competition both at home and abroad, your organization will be eaten by one who is ready!

Evaluation of Current Maintenance Practices

Maintenance Fitness Questionnaire

Consider a Meeting to discuss and learn

For maximum benefit from this list, use sections of it as an agenda for a meeting about the nature of maintenance in your facility. To use the list in this way, have each person go through the questionnaire and rate your department. Ask the people them to write notes in the margins if they have any questions about the questions. Also tell them to:

- Put a star * in front of those questions that you think are of vital interest to your organization (not all organizations need all these sys tems and procedures).
- Conduct a series of meetings with stakeholders to discuss these starred * issues.
- Discuss at least priority 1 and 2 items as well as any items with stars.
- You might focus one topic on one meeting such as one meeting on Initiation and Authorization of Work'.

The numbers in parentheses in the following paragraphs denote the priority of that question in the author's eyes.

A. Initiation and Authorization of work

(1)

1. Is a written work order on a printed form or computer-generated form (or on a laptop, tablet PC, supercharged cell phone, or PDA screen) used for all jobs?.

(2)

2. Is the written (or electronic) work order available prior to starting a job for all work except genuine emergency repairs?

3. Do all Work Orders have a unique work order number and an equipment or area ID number (called tag number, asset number, maintenance worthy item)?

4. Is a regular (weekly) coordination meeting held between operations and maintenance departments to set priorities and choose work for the next week?

5. Is a reasonable "date wanted, time wanted" space provided on each work order with restrictions against ASAP, RUSH, HOT (one plant had a rule that if a requestor checked the "HOT" box they were authorizing overtime or contractor usage from their own budget!)?

6. Is all work requested classified for repair reason (the source of the work), such as corrective, routine, breakdown, PM, diagnostics, construction, PCR (Planned Component Replacement), project, modernization, etc?

7. When quality or statistical process control limits are exceeded, is a maintenance request initiated.

(3)

8. Are all incoming work orders, work requests, and notifications screened (also may be triaged for urgency) by a single person or selected group, before issuance, except for genuine emergencies

(4)

9. Are all maintenance users with the authority to request maintenance work specifically trained in proper techniques to fill out work orders.

B. CMMS, Systems, and Procedures

(1)

1. Do all maintainable units (machines, buildings, systems, etc.) have unique ID numbers?

2. Do mechanics and supervisors have the training, time, knowledge, positive attitude, and access to the CMMS (Computerized Maintenance Management System) to investigate a problem?

3. Is all garbage and faked data kept out of the CMMS? The data output by the system must be recognized as accurate and useful by both the management and the workers.

(2)

4. Is all repair history from the date in service made available immediately, with enough accuracy to detect repeat repairs, trends, and new problems.

5. Is there a roadmap for increased use of the CMMS from wherever you are now to where you want to be?

6. Can the CMMS or manual system isolate the `bad actors' (the problem machines, craftspeople, or parts) and are there protocols in place and in use to generate exception reports to identify them?

7. Once the 'bad actors' are identified, are they analyzed and, if indicated, worked on in a reasonable period.

8. Can the CMMS give answers to most maintenance questions without the services of a programmer?

9. Is the maintenance manager part of the top level strategic planning for the future of the facility, product, or organization?

10. Is there an absolute commitment on a long-term basis to improve the quality and reduce the cost of maintenance among top management and maintenance management people?

(3)

11. Do all Work orders include costs (labor charges, parts charges, outside charges) so that intelligent decisions can be made from real numbers.

12. Can the system generate meaningful comparison data between like machines, buildings, and cost centers?

(4)

13. Are systems in place that detect craftsperson-induced problems (iatrogenic problems)? Is the percentage of rework or call back less than 3%?

14. Are there systems in place to detect deterioration in the uptime of major machines, lines, or areas?

15. Do the systems highlight when overtime is below or above predetermined set points such as 3%-9%?

16. Can the system look at component life and easily provide an analysis for MTBF (mean time between failures)?

17. Have unnecessary systems, reports and procedures, left over from when the organization was bigger or reporting requirements were different, been eliminated?

18. Do you have a written procedure for the work order system (SOP) that is reviewed, followed, and kept up to date?

19. Is the CMMS system supported by either a responsive vendor or a responsive IT department?

20. Is the CMMS system integrated with stores, purchasing, payroll, CAD/Engineering?

21. Are new capabilities being added to the system regularly?

C. Preventive\Predictive\Conditioned Based Maintenance

(1)

1. Does top management support the PM system with their attention, money, and most importantly, downtime when you need it?

2. When deficiencies are found during inspection are they written up as corrective work and completed in a reasonable time. (In other words, will production control or production give up a machine if a potential problem is detected during the PM inspection and maintenance wants to schedule a repair?)

3. Is the reduction and eventual elimination of maintenance requirements a mission of the maintenance department (is it really the mission and not just lip service)?

(2)

4. Do repeated or expensive failures automatically trigger an investigation to find the root cause and correct it?

5. Are the operators/users trained to do routine maintenance and PM where possible (this is sometimes called TPM)?

6. Was an economic analysis of each task list proving ROI (Return on Investment) performed?

7. Does each item relate to a failure mode that is expensive, dangerous, or common?

(3)

8. Are high technology inspection methods used, such as vibration analysis, ultrasonics, etc?

9. Are the high tech inspections integrated into and driven by the master PM schedule?

11. Are units kept out of the PM system (we call them bust'n'fix items –they bust, you fix) because they are in very bad shape or PM is not economically indicated?

12. Does the actual failure history impact the frequency, depth, and items on the task list?

(4)

13. Are lube routes established?

14. Do the lube people report abnormal machine conditions, and are they trained to do so?

15. Is auto lube equipment in use?

16. Is the PM system driven by measures of equipment usage such as machine hour, energy, cycles, pieces, tons, etc?

(5)

17. Are the task lists divided into interruptible and non-interruptible tasks to facilitate scheduling?

D. Planning, Scheduling and follow-up

(1)

1. Is the maintenance master schedule coordinated, reviewed and up-dated weekly, and signed by operations management and other interested parties?

2. Is the maintenance schedule linked to the master production schedule?

3. Are maintenance available hours (what's left after PM, breakdowns) well known to operations and used to schedule next week's backlog relief work?

4. Is the weekly schedule discussed after the week is over with an eye toward improving schedule compliance (not toward affixing blame)?

(2)

5. Is the backlog of jobs already planned, with materials and authorization to proceed (called the ready backlog), between 2 and 4 weeks of the crew's available hours?

6. Is work for each craftsperson planned and written on a schedule a week ahead for 75% or more of the craftspeople?

7. Is there a task and estimated time breakdown with material lists, on all planned work prior to the start of the work?

8. Are standard times or accurate job estimates established for all repetitive jobs, PM's, and routine work?

9. Are maintenance work orders prioritized in a rational way to insure that the most critical units are worked upon first?

(3)

10. Are jobs routinely completed on time, and on budget, as planned?

11. Does the mechanic show up to do a repair or PM on time (per schedule) 90% of the time and is the unit returned to service as planned on-time 90% of the time?

12. Is PM or scheduled work routinely delayed by jobs with preferential treatment (lower priority but need to be done) such as a new electrical outlet for a manager?

13. Is there an up-to-date master plan for all major jobs with important dates, durations of sub-projects, critical paths, labor and craft requirements?

14. Is the ratio between scheduled and non-scheduled work tracked, and are trends charted for at least three years.

E. Purchasing, Parts, and Stores

(1)

1. Is the storeroom well laid out with adequate space, controlled access, shelving, bins, and drawer units?

2. Can a maintenance craftsperson (or storeroom attendant) get the right part or find out that the part is not in stock in less than 10 minutes, 90% of the time?

(2)

3. Are all parts used for maintenance purposes tied to unit number(s) through the work order?.

4. Is maintenance involved in decisions to stock a new part or get rid of an existing part? Is the process easy to perform (could be harder for high value parts and easier for low value parts)?

5. Is there a periodic review of high turnover, high cost items for possible savings? Is there a review of expensive slow moving items to see if they belong in stock at all?

6. Is it easy to return excess parts?

7. Is it easy to deal with rebuildables (returning to shelf when completed, scheduling rebuild, etc?)

(3)

8. Do craftspeople or supervisors keep parts hidden around the plant to make sure they are available when needed (this situation shows that the supervisor or trades person does not trust the storeroom, usually with good reason)?

9. Is the information system support adequate for easy look-up and identification of unknown parts?

10. Does purchasing have a specialist(s) who routinely purchases maintenance parts?

11. Is there an on-going effort to use vendors as long-term partners for stocking, engineering, and problem solving?

12. Are the relationships between maintenance, purchasing, and stores good?

13. Are there established re-order points, order quantity, and safety stocks based on logical reasoning and not emotion?

14. Does the storeroom get planned jobs and PM lists early enough to prepare kits of parts in advance?

15. Is there an annual physical inventory and an annual review of parts for obsolescence, spoilage, quantity-on-hand, re-order point, lead-time, and safety stock?

16. Does your accounting system have a capital spares category for expensive, long lead time, `insurance policy,' parts?

(4)

17. When a unit is retired, are the special parts required for that unit removed from stock (scrapped, sold, returned, or reworked for use elsewhere)?

F. Budgeting, Maintenance Ratios

(1)

1. Is the budgeting process real (developed against the maintenance demand created by machines and facilities) or just some calculation against last year's budget with some padding (to be cut away by top management)?

2. Is the current budgeting process likely to serve the best overall long-term interests of the organization?

3. Is there a long term strategic maintenance plan including projected capital replacements, that is reviewed and updated at budget time by both maintenance and top managers?

(2)

5. Is up-time (or downtime) tracked for all critical machines or processes and publicized?

6. Are downtime reasons tracked and trended (is maintenance downtime separated from other sources of downtime such as model change-over or material problems)?

7. Is there a well known cost of downtime by machine, process, or facility (does everyone know what it is)?

(3)

8. Is there a maintenance improvement fund or a maintenance R & D account to charge experiments to improve maintainability?

9. Is there recognition that good maintenance practice has a major impact on other budget line items such as energy, or regulatory compliance?

10. Is the relationship between the amount of output (production quantity such as cases processed or tonnage shipped) and maintenance staffing, well understood and used for budgeting?

11. Are budget savings available for reinvestment in other cost reduction programs?

12. Is the ratio of maintenance dollars to overall revenue dollars (or some other commonly-used overall measure), tracked and trends kept for at least 3 years?

13. Do major pieces of equipment have individual repair/replace budgets?

14. Is the budget performance kept up-to-date by month with adjustments as necessary? Is the performance well publicized?

G. Guaranteed Maintainability

(1)

1. Are drawings and specifications on all new machines, processes and buildings reviewed for ease of maintainability by the maintenance department and operations early enough in the process (usually reviews are after 30%, 70% completion of engineering) so that changes can be made without adverse impact to the whole project?

(2)

2. Are specifications for product, process, or buildings, discussed with the maintenance department to see if they are in line with existing skills, parts stocked, and tools? If they are not in alignment are sufficient funds and time made available to get training and buy tools and spare parts?

(3)

3. Are new technologies on machines and building systems discussed with maintenance so that training and test equipment can be procured both before and during installation or construction?

4. Do designers or equipment buyers use failure and cost histories to make better decisions?

H. Training, Hiring, and Employee Development

(1)

1. Are 1-5% (20 to 100 hours per year) of a technician's direct hours spent in craft, multi-craft, or other related training (such as manufacturer's training, training on new technology, etc.)?

2. Is cross training a goal for the department to make it possible for craftspeople to do the `whole' job and provide scheduling flexibility?

3. When hiring, is there a consistent process to identify and hire the best maintenance workers with the final decision being made by the maintenance department?

(2)

4. Is training for continuous improvement of skills part of the mission of the department?

5. Has there been a recent assessment regarding the technology and skill level required to maintain the machinery, equipment, and processes, and a comparison with the average skill level of the employees in the maintenance department?

6. Is there on-going training of operators to improve their operation of the equipment, help them correct minor faults, and improve their powers of observation (to improve the quality and accuracy of the maintenance request)?

7. Is training tracked and reviewed by supervision, the craftspeople, and management at least quarterly?

(3)

8. Is training for the maintenance department part of all new equipment acquisition contracts?

9. Is training provided for maintenance personnel and maintenance users in filling out work order fields accurately, to facilitate future uses of the database?

Maintenance Processes

Maintenance Quality Improvement

Quality control in maintenance is hard to define. The usual definition in production is quality to consistently produce parts with low variation. Maintenance quality usually deals with the effects, not the repair itself.

In some circumstances	maintenance quality might	=	Reduced downtime
In others:	maintenance quality	=	Reduced scrap
	maintenance quality	=	Faster start-up
	maintenance quality	=	Quicker response
	maintenance quality	=	No repeat repairs
	maintenance quality	=	Keep production within specification
	maintenance quality	=	No interruptions
	maintenance quality	=	Satisfied user

Maintenance is a business with business processes

To improve maintenance quality we must attack defects in the processes that feed maintenance. The processes of maintenance to a large part determine the outcome of the maintenance service:

——PROCESSES——-

Project	Engineering	Trades people	Customer Satisfaction
Breakdown	Vendors		Production
Initiation	Planning	Execution	Warehousing
PM	Part vendors	Outside shops	Administration
Corrective		Outsourcing	Contractors

→

SALES	SUPPLIERS	DELIVERY OF PRODUCT OR SERVICE

Maintenance is a business. In every business, staying close to the customer helps ensure survival and allows the business to thrive. Quality as a measured value comes from a subtle understanding of the customer's needs.

Traditionally, the mission of maintenance is to support the productive output or activity of the organization. At one time that mission was accomplished by a single-minded focus on the machines, systems, and buildings (called assets). This paradigm is best stated by the dictionary definition of maintenance management that is "the act of directing the preservation of assets."

There is a new view in organizations, a new paradigm for the whole business, that refocuses on the process and the people rather than the product. The new view causes a revolutionary shift in the mission of maintenance.

The new paradigm of business is to focus on streamlining, reducing the inputs into, making more responsive the process, of manufacturing. As a result, the mission of maintenance must work toward the continuous improvement of all processes of the manufacturer. The new mission includes the idea that the maintenance department should work endlessly to reduce and where possible eliminate the need for maintenance. The ultimate expression of quality is not quick, high-quality repairs but rather no repairs needed at all. Coupled with this goal is the parallel goal of maximizing the useful output of a given set of assets.

The new mission serves both classes of customers powerfully. The customers are denoted by the big `C' (ultimate customer who is outside the organization) and the little `c' (internal customers). The sale of our product to the `C' creates an invoice, which results in a payment that pays our salary. The little `c's' are the internal customers such as production, finishing, assembly, administrative offices, warehousing, etc. Most maintenance departments spend all of their energy thinking in terms of the small `c' and ignoring the needs of the big `C'.

Any activity that has an impact on the big `C' such as breakdowns, downtime, and quality variation or missed delivery is critical and has to be looked at very closely.

Initiation → Planning → Execution → QC → Customer Satisfaction

Process of Maintenance

This book focuses on the process of maintenance in the middle of this diagram. The processes of maintenance to a large part determine the outcome of the maintenance service.

Initiation: The process starts with a need. The need could be for a PM Service (initiated by maintenance CMMS) or to counter deterioration in a physical asset. A PM inspector, maintenance person, or an operator might notice the deterioration. The process is optimized to simplify notification of the maintenance department and collect all the relevant information at the lowest cost.

The process continues with the planning function, material ordering, prioritization, and scheduling of the work needed. The first step is to triage incoming work requests to separate out emergencies. Work that can wait is formally planned to optimize the use of scarce resources. Proper planning also studies and manages the hazards introduced by the job to the mechanic, to other employees, and to the environment. Scheduling insures that the person, part, tool, permission, information, and the asset to be serviced converge at the right place at the same time.

The execution process can be as simple as giving the planned job package to the mechanic or having a supervisor, fore-person or gang leader issue the work order. Execution is tactical. All the problems that crop up, and all the gaps between the plan and reality now have to be dealt with.

QC in maintenance

The maintenance QC process insures that the right job is done on the right asset in the right way with the right materials and tools. Of course, if quality is not built in to the plan it is too late now to fix it. All we can do now is rework (the more expensive option). In many workplaces the mechanic him/herself serves this function.

For the customer to be satisfied the process is also responsible for reporting to the customer that the job is complete.

An unsatisfactory process is one that depends on the individual heroism of maintenance management, support staff and technicians to serve the customer and to produce quality work. For example, the stores group is a vendor to the maintenance process. If a part that should have been in stock is not in stock then service to the customer suffers, excessive downtime might result, and a delivery window might be missed. Individual heroism such as digging through the dumpster and re-machining a discarded part to fit might be the only way to serve the customer. Good service with a defective process is the exception, and can only result from special effort.

On the other hand, a good process provides good service with only occasional acts of heroism. High quality lies at the core of the process. Improvements in the quality of maintenance delivered, envelop the entire process from improved supplier relationships, through improved work order handling, to better understanding of true customer needs.

Quality equation

In a production environment:

Improvements in maintenance quality = *lower unit costs*

Because when you improve quality then:

Decreasing costs result from less: waste, scrap rework, maintenance rework, downtime, and variations, which improves productivity while reducing disruption.

Quality in maintenance makes the organization more competitive and allows it to capture greater market share with better quality and lower cost (and the greater margins contribute to the company becoming the lowest cost producer).

As an added bonus to you and your maintenance staff, your company gets to stay in business and provide steady jobs

Every maintenance operation should define quality in a way to be useful to the operating environment.

The late W.E Deming was considered the chief quality guru in the last generation of quality experts. In fact, the quality award in Japan today is the Deming award. Deming had much to say about quality in manufacturing.

Based on information provided by the W. Edwards Deming Institute, Dr. Deming (William Edwards Deming) was born in Sioux City, Iowa on October 14, 1900 to William Albert Deming and Pluma Irene Edwards. In 1917, he enrolled in the University of Wyoming at Laramie. In 1921 he graduated with a B.S. in electrical engineering. In 1925, he received an M.S. from the University of Colorado and in 1928, a Ph.D. from Yale University. Both graduate degrees were in mathematics and mathematical physics. Dr. Deming, surrounded by his family, died at his home on December 20, 1993

From his work experience it did not seem likely that Dr. Deming would become the world leader in quality management. In Deming's own words, from his resume in 1974, describing his experience after getting his PhD from Yale: "I entered service in the Department of Agriculture in Washington for work in mathematics and in mathematical statistics, and was there until 1939, when I transferred to the Bureau of the Census to work on problems of sampling.

During the years 1930 through 1944 I lectured twice weekly in advanced mathematics at the National Bureau of Standards. I instituted in 1935 a program for the teaching of modern theory of sampling at the Graduate School in the Department of Agriculture in Washington, and I continued in charge of the courses in mathematics and statistics there for 20 years.

I was Mathematical Advisor to the Bureau of the Census from 1939 through 1945, where I took part in the development of sampling procedures that are now known and used all over the world for current information on employment, housing, trade, and production. People come from government offices and from industry in many parts of the world to work in our Census, and to study these methods....

From 1939 to 1942 I worked with the War Department (now the Department of Defense), to assist the development of procedures for the statistical control of quality, to improve the precision and dependability of manufactured products. The American Standards Association published this work in three standards, which were later adopted by standardizing bodies in several other countries."

The work for the War Department had Deming apply his vast knowledge of statistics to the problems of production and to the quality of the military hardware being produced. He learned lessons about the shop floor management of product quality. One of the most important lessons he learned came after the war was won.

Deming (and a group of like-minded mathematicians and statisticians) thought that the gains in quality control during the war would be applied to civilian production. That did not happen. As soon as civilians came back from the war they reverted to the old methods and virtually wiped the slate clean. All the advances were abandoned.

Deming was sent to Japan as part of the Marshall Plan in 1949. He made 11 trips to Japan through 1974. His mission was to help get Japan back on its feet after the bombardment of World War II wiped out its manufacturing base. He was greeted warmly and he was listened to closely. The Japanese realized that they could not compete on price alone because they had few natural resources. Quality was going to be the ticket to the markets. Japanese manufacturers realized that Deming's methods for quality management were essential steps back to self-sufficiency and prosperity.

Deming designed a seminar on a statistical approach to quality. His first presentation, in 1950 was attended by the presidents and other chief executives of the major Japanese companies. He gradually realized that the impetus for quality must come from the top and the top people must be trained to understand basic statistics and quality concepts.

The surprise is that Deming's points apply to maintenance also.

W. Edwards Deming's Fourteen Points were first discussed in his quality seminar in 1950! The following list is adapted (without his knowledge or permission, I might add) to the realities of maintenance management in a modern manufacturing plant.

1. Create constancy of purpose toward improvement of products and services with the aim of staying competitive, staying in business and providing stable employment. Maintenance deterioration usually takes a long time. Any effective maintenance strategy must also have a long horizon. Resources must be allocated for good maintenance practice and not taken away with every bump in the quarterly results.
Deming's deadly disease related to manufacturing was the lack of constancy of purpose to plan products and services that will have a market, keep a company in business, and provide jobs. Maintenance issues (like the wearing out and failure of a compressor or boiler) take a long term to develop. Only an equally long-term view will be effective. A moving agenda for the goals of maintenance will work against the department.

2. Adopt the new philosophy. Awaken to the challenge. Take responsibility for and leadership in change. Our maintenance departments often are the last areas of the organization to realize the need for change. The department needs to be dragged, kicking and screaming, into the new corporate culture. Looking toward the future, I see a maintenance department providing leadership for the rest of the organization. Nowhere else is high quality so closely related to safety, and high self esteem. After all, quality is intertwined with the very history and culture of the crafts.

3. Cease dependence on inspection to achieve quality. Build quality in. Quality comes from skilled and knowledgeable mechanics given good tools, reasonable work conditions (free from the fear of bodily harm), adequate materials and enough time to do the job.
Quality comes from choosing well-designed equipment that does not need much maintenance. What maintenance the equipment does need must be accessible and easy to perform. Quality comes from pride in a job well done. Good Maintenance management acts decisively to make sure craftspersons can complete jobs they start so that they get the feeling of satisfaction from jobs well done.
Management leads by example with ceaseless training, coaching, and systems analysis. When defects occur, management concentrates on the system that delivered the defect rather than having a preoccupation with finger pointing.

4. End the practice of awarding business based on price alone. Instead, minimize total cost. Move toward a single source for each item and on a long-term relationship of loyalty and trust. A revolution in purchasing is at hand. More and more organizations are looking at the total costs of a part or the life cycle cost of a machine. Some economies are false, and hurt the overall goals of the organization. A low cost bearing might be the most expensive bearing you ever buy.

5. Be dissatisfied with your current level of maintenance management abilities. Improve the system of production and maintenance constantly and permanently, to raise quality and productivity, and thus constantly reduce costs. In today's market, the way things used to be done is never going to be good enough for the future. All improvements and growth flow from dissatisfaction with the status quo. Measurement must be built into the maintenance information system and there must be continuous efforts to improve both the visible and the invisible performance.

6. Institute training on the job. Training should be mandatory for mechanics the way continuing education is for doctors or teachers. Our factories and facilities have today's levels of technology and too many of our maintenance people have yesterday's skill sets. To maintain effectiveness we must train to bridge the gap. Special incentives should be given to the people on your staff who deliver on-the-job training. These informal trainers need instruction in how to teach adults. They also need back-up materials and support to deliver the best possible training.

7. Institute leadership. The aim of supervision should be to help people and machines do a better job. Supervisors should serve their subordinates by removing impediments to productivity. Supervisors should insure that the mechanic, the tools, the parts, and the unit to be serviced, converge at the same time. The supervisor should also be the lightning rod for disruptions from management and production. Unless there is an emergency, the mechanic will not be disturbed, because interruptions reduce quality and worker satisfaction.

8. Drive out fear, so everyone may work effectively for the company. Fear of the loss of a job or of dangerous conditions interferes with the mechanic's ability to concentrate. Fear gets in the way of the pride a mechanic feels in a job well done.

9. Break down the barriers between departments. Everyone's expertise is needed for constant improvement. With scarce resources we must include knowledge from other departments and groups to come up with the best overall solution for the organization. Maintenance problems can become complex quickly, with financial, marketing, purchasing, quality, and engineering ramifications. The best solution to a problem might not be the best maintenance solution (such as run until destruction to fill a critical order). Information for the best solution might come from another department and another expertise.

10. Eliminate slogans, exhortations, and targets for the work force asking for zero defects, and new levels of production. Such exhortations create adversary relationships. The bulk of the problems related to quality and production belong to the system not the people. Stable processes create quality. Create stable processes, producing quality outputs, and the people will feel the way the slogan speaks without coercions and alienation.

11. Eliminate work standards, quotas, and management by objectives (MBO). Work standards and quotas are associated with management styles that treat the maintenance worker as someone needing to be told exactly what to do and how long to take. Standards are useful for scheduling and to communicate management's expectations. It is difficult to not use them as a production whip. Such usage is a disaster in maintenance situations because we want the mechanic to take the time needed to fix everything they see (within reason!), not just the original job. We must trust the mechanic to look out for our interests, particularly when we are not there. The problem with MBO is that it focuses on visible, measurable, aspects of maintenance. Many of the real issues of maintenance concern aspects of the environment that are hard to measure.

12. Remove the barriers that rob the worker/engineer of his/her right to pride of workmanship. The responsibility of supervisors must be shifted from numbers to quality and improvement values. Tradespeople must be allowed to feel pride in their jobs when they are well done. Maintenance managers and supervisors must not allow anything to stand in the way of that pride.

13. Institute a vigorous program of education and self-improvement World-class maintenance departments make a commitment to invest 1-3% of their hours in training for all maintenance workers. Technologies are changing and skills must change too. A world class automobile manufacturer mandates 96 hours of training per year for everyone. A high tech manufacturer requires 110 hours. A flexible and highly-productive department is where people are trained and can shift from trade to trade, maintenance, to construction, to production, based on the needs of the day.

14. Put everyone in the organization to work to accomplish the transformation. This transformation is everyone's job, and it requires the talents of all the employees. It requires all the talents of each person. When a hotel chain had the housekeepers meet with the architects (for a new hotel) the result was concrete suggestions to improve the designs that reduced maintenance costs and improved the rooms for the customers.

Along with the 14 points were problems or barriers to effectiveness. Deming noted that there were obstacles that stood square in the way of success. He called these deadly diseases.

Deadly diseases and obstacles to success

1. The supposition is that solving problems, automation, gadgets, and new machinery will transform industry. Maintenance problems are people problems. The systems, attitudes, and approaches are at issue. For example, the attitude of maintenance as a necessary evil, or of maintenance workers as grease monkey slobs, must be transformed. The transformation starts in the minds and hearts of the maintenance department and then flows to the rest of the organization.

2. Emphasis on short-term profits, and short-term thinking feed on fear of unfriendly takeovers, and with a push from bankers, over-zealous managers concentrate on lining their pockets with options and providing owners with dividends. In these conditions, top management will squeeze maintenance to reduce costs below the level that is necessary to avoid deterioration. The cost reduction is temporary, the asset will deteriorate and long-term integrity of the process will be compromised. Maintenance requires long term planning and commitment.

3. Evaluation of performance, merit rating, or annual review. The question about annual reviews and performance ratings is, what useful outcome flows from these procedures? In most instances, production of mechanics is more related to how much management gets in their way rather than to their personal qualities. According to Deming, annual reviews rarely change behavior.

4. Mobility of managers and job-hopping. In one major beverage bottling plant, the average tenure of the maintenance manager was 22 months. Some lasted as few as 9 months. All the managers came with bright ideas and wanted to prove themselves. The result was a complete lack of focus on long-term goals and plans. As each manager tried to cut costs and shine, the negative results impact fell to the next player. This job hopping in management without a master plan dramatically exacerbates the short-term view.

5. Management by use of only visible figures, with little or no consideration of figures that are unknown or unknowable. For example, when you invest in training for your maintenance crew where does the increased asset show up? When, after spending $100's of thousands in a long expensive trial and error development process, a firm finally develops expertise in a new process, this expertise, this new asset, does not appear on the balance sheet. It is important to measure. But to realize that much of what goes on in maintenance is unknowable (or at least difficult to measure).

6. Hope for instant pudding. Changes to fundamental processes take time. In the current US culture it is hard to imagine instituting a change in processes that could take 5 or 6 years to complete. In fact, if you start with a typical reactive mainte-

nance department it could take 5 years or more to create a proactive department with a TPM type partnership in maintenance and production.

7. Search for examples. We often think that if something worked in another machine shop or foundry it will work in ours, but maintenance in factories has no strict rules, so examples from another plant even in our industry may not be useful or even relevant.

8. "Our problems are different." Actually many people's problems are the same. In the PM area, although no two plants will have the same exact schedule, many of the problems will be the same. In our public sessions, maintenance managers in widely different industries, sizes, and sophistication, often marvel at the similarity of the problems.

9. Poor teaching of statistical methods in industry. Industry is just waking up to the value of statistical methods of explaining what happens in the shop. Application of simple statistics to PM or PCR intervals would improve effectiveness. Simple relationships such as failures to PM would show the effectiveness of the frequency selected. Statistics replaces seat of pants reasoning, panic logic, and historical prejudices, with testable and verifiable conclusions.

10. "Our trouble lies entirely within the work force." Deming stresses that any production system that is a stable system will produce a certain number of defects. Changes in the work force are irrelevant to the output. Only changes to the system can have an impact.

11. False starts with inadequate planning, top-level support, and lack of follow through, kill quality improvement in most places. Serious thought and planning are needed before starting. Commitment must start in the highest levels in the organization. Buy-in at each level must be earned, worked for, and appreciated, before proceeding to the next level.

12. "We installed quality control." But quality control is a way of life. It is a daily diet. You do not install it, you become it. This axiom is also true for safety, and environmental compliance. These attributes must be planned into every job, not built on at the end!

13. The unmanned computer is one of the dangers of wholesale computerization of maintenance. The computer is a great tool that, like any great tool, is frequently misapplied. Allow the people to have their say, and make sure the computer answers to some one (a real person) and that they can overrule the machine.

14. The supposition that it is only necessary to meet specifications. Many of the important aspects of a component are not included in the specifications. You don't know which attributes are important until you try changing vendors and find out that your entire process depends on qualities of a particular vendor's products that are not covered by the specifications.

15. The fallacy of zero defects. Every system produces defects. Ultra high quality requires enormous sample universes to establish the defect rate. In maintenance, expect defects and build systems that uncover and deal with them. It is also essential that we use defects and mistakes as teaching aids.

16. Inadequate testing of prototypes. By starting manufacturing on inadequately tested prototypes we strain the system of improvements. There will be so much ground to cover before everything stabilizes that the product will be half-baked for a long time. To leapfrog this phase, exhaustive testing should be built in. The same is true for new machines and proper commissioning.

17. "Anyone that comes to try to help us must understand all about our business." The sad truth is that if the solution to your problem was commonly known in your industry you would probably already know what to do.

Maintenance Information Flow

Where does your work come from?

Several sources are internal (PM tasks, corrective maintenance), and the rest are external to the maintenance department. A study of your information flow might show where requests for work could fall through the cracks. It would also show where you are not capturing the minimum amount of information to dispatch maintenance workers efficiently.

In one maintenance department, the work order went through 43 steps from conception to completion and filing. Business process review reduced those 43 steps to 23 (which are still quite a few). The business process of accepting work is an important one. We must process the jobs with as little overhead as possible, but make it easy for the customer. The big but is that the process should yield all essential information about maintenance activity for future analysis.

All requests for service should converge in the maintenance control center and be reviewed by a single person or group. If you use a team structure, then the team point person on planning and scheduling should review the requests. Emergencies should be handled directly. The goal of the system is to serve the customer efficiently with minimum overhead while maintaining control.

Maintenance information flow is independent of the size of the factory. Large organizations might have whole departments providing each function. Small ones can literally use four clipboards for all maintenance activity.

1. The wide-ranging funnel is the in-box. All services requested from all sources go through the In-Box. including incoming E-mail, phone messages, notes, and radio calls, from all maintenance users, PM tickets from the CMMS itself, and project packages from engineering (if the labor is to be provided by maintenance). Time/date stamping should take place when a job is placed in the In-Box. The In-Box should be scanned frequently to keep tabs on high priority work that no one mentioned.

The first step is triage. Emergency jobs are dispatched at once to a floor supervisor or emergency crew lead person. In some systems, the time when this emergency job was dispatched is recorded. After the true emergencies are processed, the remaining jobs are reviewed.

Incoming work requests are taken from the `In-Box' and are reviewed for authorization, and intention to complete (if there is no intention then the job may be discarded or moved directly to the wish list). The In-Box is the conduit into the backlog. The In-Box should be emptied reasonably quickly.

2. The next stop is backlog. The backlog is the holding tank that helps regularize the flow of work to the maintenance shop. Jobs may enter the backlog in clumps but they are parceled out in a smooth flow that is directly related to the hours available. Emergency jobs pass directly to the field so are not considered to be backlog. By custom, PM jobs are handled outside backlog (but of course, are included in the calculation for how many hours are available for backlog relief).

There are several types of backlog. These types of backlog correspond to the status of the job.

Total Backlog: All backlogs are part of the total backlog	Backlog waiting for materials Backlog waiting for authorization or approval Backlog waiting for engineering Planning or planner's backlog (jobs waiting for planning) Backlog waiting for weekend or shutdown Ready backlog Other

Types of Backlog - Status

Some computer systems (CMMS) maintain status codes for each job. These status codes follow the job through all the steps before the job is released to the shop. Status codes could include: Waiting to be planned, Waiting for Engineering, Waiting for P.O. to be issued, Waiting for Material to be received, Waiting for weekend (night or summer), and Waiting until shutdown, or Ready for fill-in assignment. These status codes denote the sub-category of backlog.

The two most important types of backlog for this discussion are the total backlog and the ready backlog. Most jobs from the In-Box go directly into the total backlog. At that time engineering, parts ordering, and planning are ini-

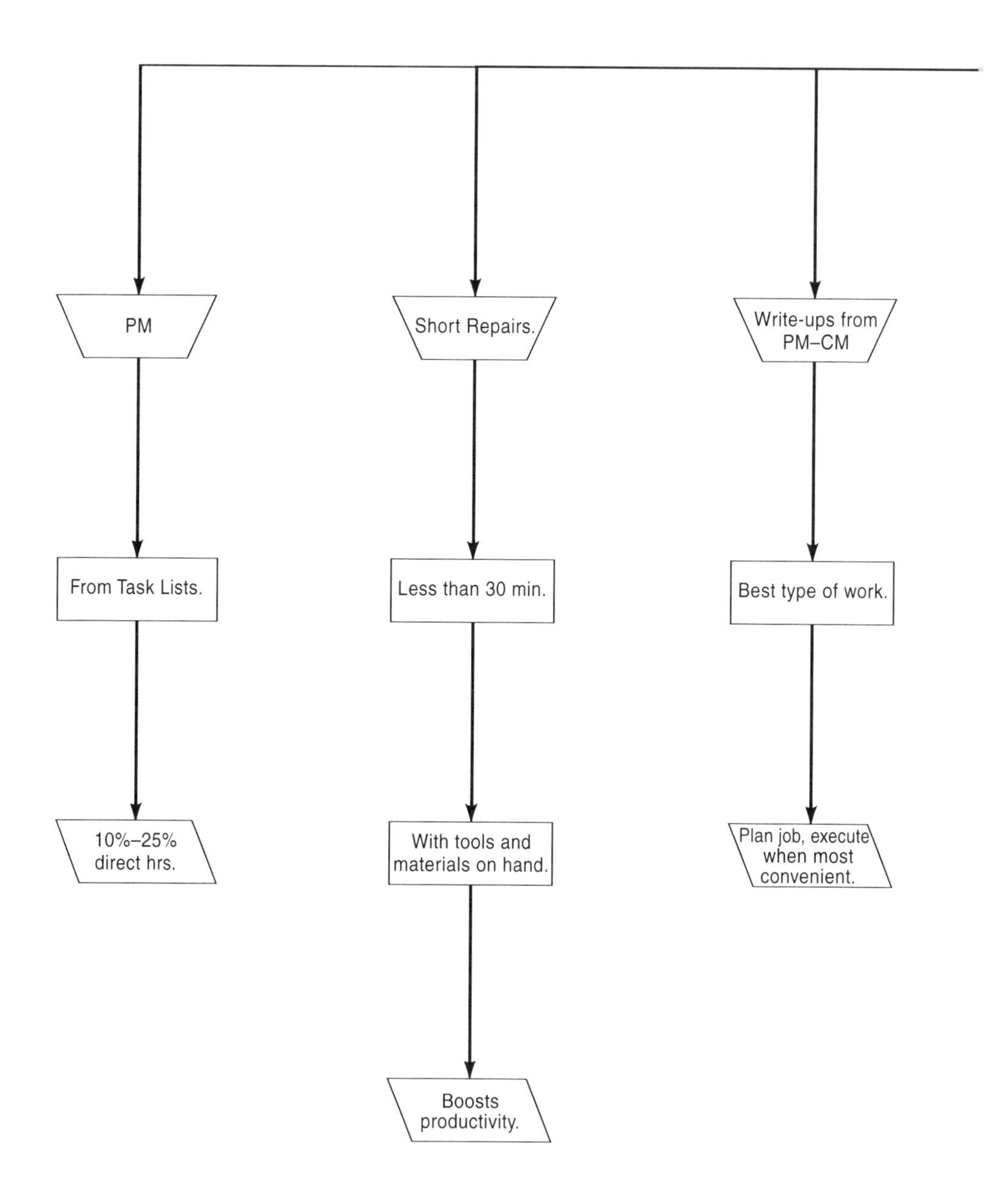

Sources of Maintenance Work

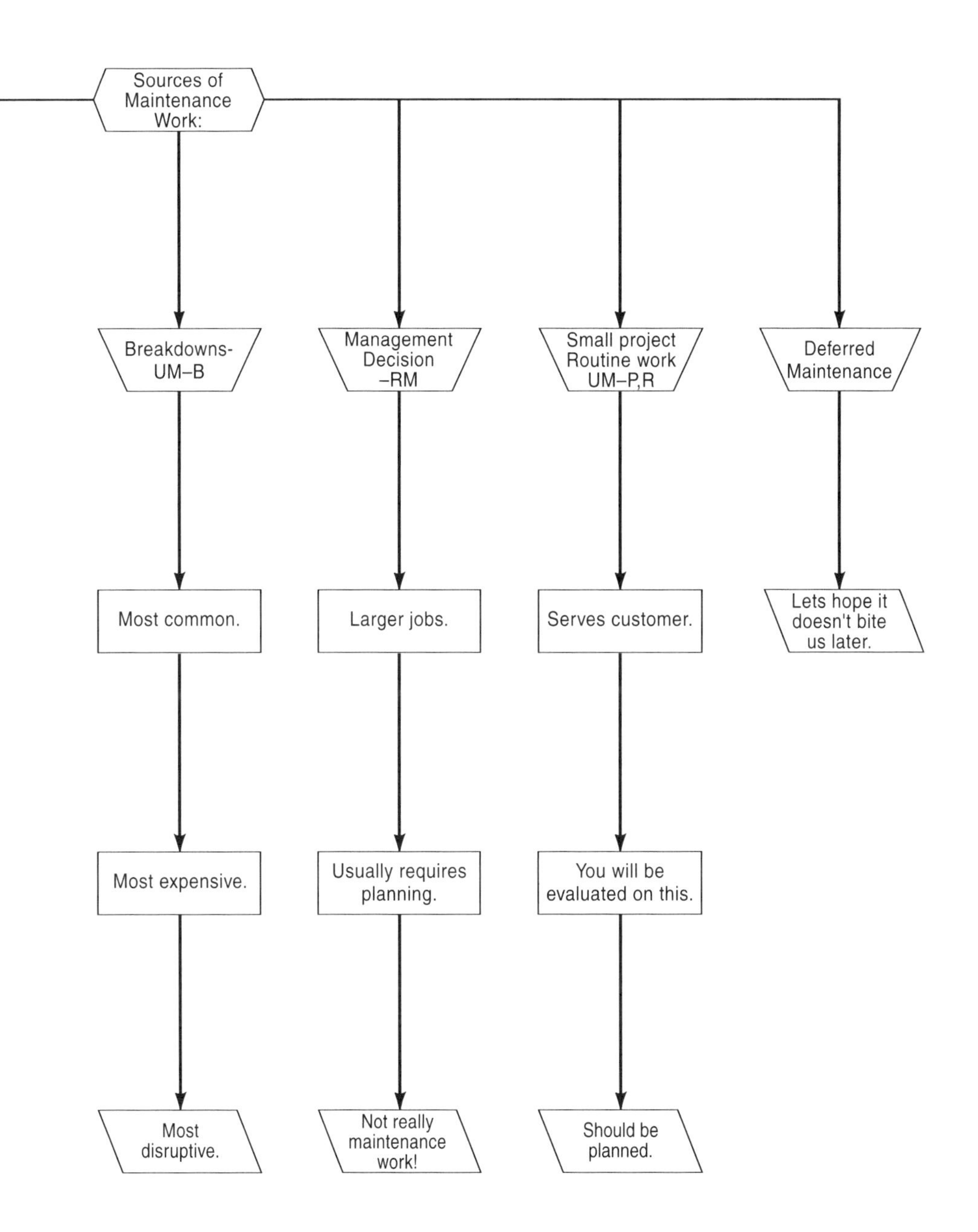
Sources of Maintenance Work:
Breakdowns- UM–B
Management Decision –RM
Small project Routine work UM–P,R
Deferred Maintenance
Most common.
Larger jobs.
Serves customer.
Lets hope it doesn't bite us later.
Most expensive.
Usually requires planning.
You will be evaluated on this.
Most disruptive.
Not really maintenance work!
Should be planned.

tiated. The work is scoped out (job site might be visited and the job investigated). An investigation might uncover that the job has been done before and a job plan or engineering drawings might already exist.

These activities are planning functions (whether there is a planner or not). Planning might be as easy as making sure that the materials for a machine repair are in stock, or as complex as coordination of contractors, parts, and rental equipment for a major installation. When planning is complete, materials are in stock, authorization is obtained, and any barriers to production are handled, the job status changes to ready and the work order joins the ready backlog.

The exception are jobs that go directly from the In-Box to the wish list (where there is no near term intention of doing the indicated work). Work orders that you have decided to defer might be filed in the organization's files of wish list jobs against the time when some resources are freed up. In either event, the work flows out of the holding area of the in-box totally bypassing the backlog. Jobs requiring shutdowns are stored in the 'waiting for shutdown' file which may or may not be reflected in the backlog numbers.

After the job is planned and resources are secured it moves to ready backlog. All ready backlog jobs have been authorized, parts are available, priority has been set and planning (if required) has been done. Managing the backlog is an important way to manage an entire maintenance department. It is essential that the ready backlog not fall below two or three weeks nor increase to greater than 4 weeks per tradesperson. Too low a backlog will encourage the trades people to stretch out the work and to work inefficiently to avoid layoff. Too high a backlog will cause the customers to suffer unreasonable delays for routine work requests and encourages maintenance managers to jump trades people from job to job without completing them just to shut customers up.

3. Once the job is scheduled and issued to the supervisor, it then moves to the open or pending status. By convention, the job is still formally on the backlog list until it is completed. In some CMMS the estimated hours for the job are reduced as work is done, and data is entered until the hours worked are equal to the estimate. At that time the job is still in backlog (until it is completed and closed) but without any hours. Jobs that are scheduled are jointly chosen by operations and maintenance for the next schedule period.

Jobs with an open status or that are on the `Pending' clipboard, have been scheduled and should be completed in the schedule period (generally a week). The weekly schedule includes all the jobs from backlog that both the operations department and the maintenance department agree to be done.

This schedule or clipboard should be reviewed regularly for jobs that get stuck. A stuck job that stays stuck is a problem for the manager, the department and the user. Extra effort (sometimes an outside contractor), is needed to unstick these jobs.

4. When jobs are completed the tradesperson records information about the repair, check, or intervention. It is important to know precisely what was done, how long it took, what other resources were used, and any ideas or comments. The job paperwork (usually referred to as the work order), once completed, is entered into the CMMS (if there is one) where it is costed and posted to the various files.

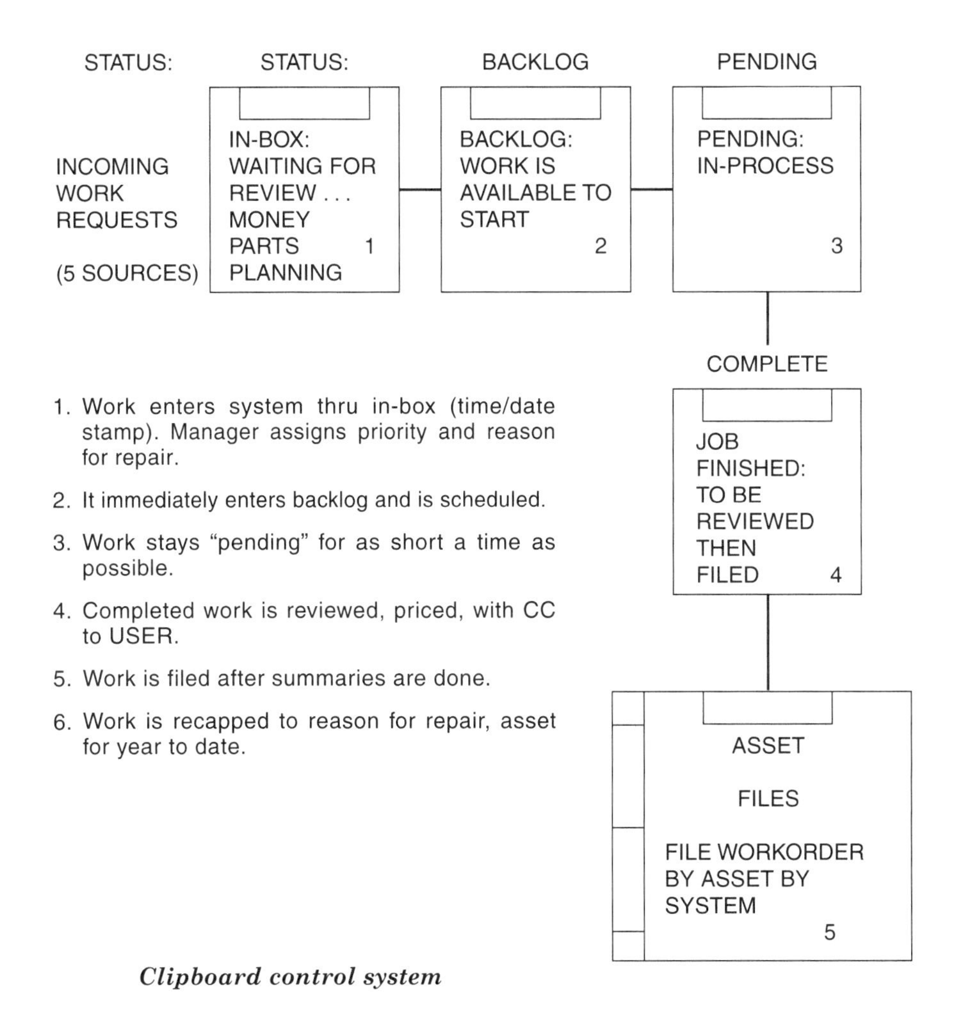
STATUS:
STATUS:
BACKLOG
PENDING
INCOMING WORK REQUESTS
(5 SOURCES)
IN-BOX: WAITING FOR REVIEW . . . MONEY PARTS PLANNING 1
BACKLOG: WORK IS AVAILABLE TO START 2
PENDING: IN-PROCESS 3
COMPLETE
JOB FINISHED: TO BE REVIEWED THEN FILED 4
ASSET
FILES
FILE WORKORDER BY ASSET BY SYSTEM 5
1. Work enters system thru in-box (time/date stamp). Manager assigns priority and reason for repair.
2. It immediately enters backlog and is scheduled.
3. Work stays "pending" for as short a time as possible.
4. Completed work is reviewed, priced, with CC to USER.
5. Work is filed after summaries are done.
6. Work is recapped to reason for repair, asset for year to date.

Clipboard control system

If you are operating manually then every work order should be costed and then entered on to re-cap sheets for the manual Asset File System.

In both instances, all sketches, photographs, parts lists, warranties and little booklets (that come with thermocouple, PLC modules, pumps, etc.), go in the file. In some modern systems, sketches and data sheets can be scanned and attached to the work order electronically. Costing, research, leg work, economic analysis, and root failure analysis notes, also go into the file in the appropriate section.

Typical Process Benchmarks

1. Respond to 99% of emergency service calls within 2 hours. Repair if possible and carry to pending state if not. Secure building, machine, and user, before leaving site (even if the item cannot be repaired immediately).

2. Respond to 99% of routine service calls within 4 days. Carry to pending state. Complete all work started if possible. Secure machine, building, and user.

3. Clear open jobs (Pending clipboard) weekly.

Managing Maintenance Through Planning and Scheduling

Maintenance Crewing

One important question is how should we organize the maintenance staff most effectively for all the types of work? The three types of work we recognize are PM/PdM, Emergency, and backlog relief. These work types require slightly different skill sets, and very different mind sets.

Maintenance departments with only a few tradespeople will scoff at the notion of making their 5 or 7 maintenance workers into three crews. The truth is that you probably already informally divide them. You could take this opportunity to make the assignment more formal. Along with the formality is the idea that each type of work has slightly different training requirements, and different tooling requirements.

Intelligent crewing decisions accept the current reality of the kinds and amounts of work that come up today. Crewing decisions also merge people's personalities and skills with these kinds of work.

Managing maintenance through planning, scheduling (for a complete reference on this topic consult Maintenance Planning, Scheduling and Coordination by the author and Don Nyman published by Industrial Press)

Jobs enter the backlog to be planned and leave the backlog when they are completed. In between entering backlog and completion there, is first job planning, then choosing the job for the work list, and finally, weekly scheduling. Planning and scheduling are different unique tasks that take different skills. There is frequently some confusion between the functions of scheduling and the functions of planning. Part of the problem is that in most maintenance departments the two functions are compressed together, particularly on smaller jobs. The easiest way to differentiate between scheduling and planning is to think of a military battle.

Military planners look at several probable scopes of the battle, lay out the battle steps (where troops and weapons are concentrated at a given time), pre

Name of crew	Possible percentage of hours	Crew mission
PM and PdM crew	20%	This is the crew for all PM and PdM activity. It focuses on PM and is the last to be interrupted when break downs pile up. In this scenario, the PM crew has a week of PM ahead of it and is involved in machine assessment and maintenance improvement. Of the three crews, this group uses tools the least and their minds the most. The work is the most repetitive of the three groups, and requires mechanics who can work without supervision and like to follow lists and write things down. They might not be your 'best' mechanics but they are by far the best analysts and observers. They pride themselves on seeing subtle signs of deterioration and when everyone disagrees, being right about it.
Emergency crew	20%	This crew is available to fix anything on a moment's notice. The workers are multi-skilled, quick-witted, fully-tooled, and equipped with quick-fix technology. These are the people with crash carts equipped with a little of everything. People in this crew love the unknown. They enjoy mixing it up with all sorts of problems and they thrive on variety. Since they may not be great at following the rules, they have to be watched for safety compliance. This group might be the best mechanics (they get a lot of practice), but not necessarily. They are the fastest mechanics and thinkers. They pride themselves on fixing problems under duress and on their ability to work under adverse conditions.
Backlog relief	60%	The bulk of the backlog relief crew concentrates on doing the scheduled corrective jobs uncovered by the PM crew. These people have attributes of both the other crews. They might be the best mechanics or not but they are methodical and pride themselves on jobs well done. They also pride themselves on getting a reasonable amount of work and getting it done in the allowed time. They can work without as close supervision as the emergency crew but are not as independent as the PM crew. The backlog relief crew has access to all tools and knows the correct ways to use them.

Maintenance Crew Types

dict the duration of the engagement and the number of troops needed, decide on the best equipment to use, and calculate the logistical support for the troops and weapons. The military planner will also look at the command and control issues (battlefield communications and supervision). In reality, most plans developed sit on the shelf as a package awaiting a `go' from the commanders.

The battle plan is developed independent of the execution of the battle. In fact, there are hundreds of battle plans sitting on a shelf somewhere (probably on a computer). Nothing has been ordered, deployed or moved in the real world for any of these plans. The maintenance plan resembles the battle plan in that, until needed, it is sitting in a folder or on a disk drive. It lists all the resources needed, timing, approvals, material/part/supply lists, tooling requirements, access requirements, time, skills, any special information or requirements. While the plan is sitting there, you might never do that job again. The plan still has some value particularly if the job is similar to one coming up.

In the military, getting a 'go' from the commanders begins the execution phase of the plan. This phase might predate the actual battle by a year or more. Troops, ships, ordnance, aircraft, support units, logistical units, and intelligence, are scheduled into a theater of operations. Food and supplies are ordered. Spare parts and reserve troops are put in place. Purchase orders are prepared for parts, ordnance, fuel, and food, to fill up the pipeline emptied by what should be used. The battle plan is revised based on changes from intelligence gathered, and it is turned over to battlefield managers. Fundamentally the success of the battle depends on the quality of the initial plan. Although many past battles were saved by individual valor and genius, no one would dispute that the right plan makes it easier.

The same things are true in the maintenance world. Individual heroism or genius might save a bad plan and it's still easier to work from even an adequate plan. When we are ready to start the maintenance job (the battle), we would say that in this phase the planning is over and the scheduling, coordinating, and management phases have begun.

Want is a planned environment?

There are six elements to a planned maintenance environment. In addition to the six elements it sure helps if the bulk of your workload comes from PM work orders, corrective work orders, and small plannable projects. If the bulk of your workload shows up on your doorstep every morning, then the impact from effective planning and scheduling is going to be compromised.

1. Assign very preliminary estimates to all jobs in backlog. If you have time, make complete plans for all PM activity and the most common corrective jobs.
2. Develop a maintenance program to outline hours available for each category of work.
3. Meet with operations to choose jobs to be done next week within the produc tion schedule (sometimes called coordination)
4. Plan the individual jobs that were chosen
5. Prepare a weekly schedule outlining what jobs are scheduled and when they will be done.
6. Measure what happened and report back

Element 1 of a planned work environment: Preliminary estimates

To produce a coherent picture of the backlog, each job has to have a preliminary estimate. If no estimating has been done up to this point, there are a couple of methods to consider. Material here has been adapted from Managing Maintenance Shutdowns and Outages by Joel Levitt, published by Industrial Press.

Guess, SWAG

Although they may be educated guesses, most installations begin with guesses. An estimate (one step up from here) is based on some judgment or other information and a guess is based on nothing. Guesses are the least costly, least time consuming, and least accurate.

The disadvantage of guesses is that they are personal. They reflect personal opinion and experience of a particular supervisor, planner, or technician. Different people guess differently. Guesses are tough to tell from quick estimates. One person might be basing the guess on a lifetime of knowledge (making it more of an estimate), and another bases it on nothing.

If the planning system has feedback (where the planner gets to know quickly how long the job actually took compared with their guess) then the guesses would improve and become estimates.

For detailed estimates, one of the following methods is preferred. Estimates

Skilled planners and tradespeople can look at a job, mentally compare it with jobs accomplished in the past, and come up with an estimate. We make estimates all the time. Estimates are different from guesses in that they are based on experience. They suffer from the same problem as guesses in that they vary in accuracy with the experience and skills of the estimator.

Great mechanics often make terrible estimators. The skill to do work does not always translate to the skill to estimate the time needed to do the job. There are several aids to estimating that improve accuracy including slotting, and analytical estimating tables.

Analytical Estimates

Analytical estimates look at the pure work content for each job and add factors for travel, fatigue, and working conditions. There are tables for travel (how long it takes to go 100 feet 500 feet, etc). The number of trips for parts is based on the length of time the job is estimated to take. Factors are added in for fatigue (after so many hours), extreme heat or cold, working near energized equipment, working in confined spaces and for working at heights (add 15% for above 12 feet.)

Analytical estimation is moderately time consuming but results in excellent estimates. If an organization had an effective analytical estimation program in force for a year or more, their labor libraries would be just about full. With an effective planner's librtary going new work requiring planning might be only 10-15% of the planner's workload. Consult Maintenance Planning, Scheduling and Coordination published by Industrial Press for details on how to use analytical estimating.

Job	Description	Slot
Changing a 110V outlet or switch	No problems, wires adequate length	15 min.
Removing and replacing a fractionalhorsepower motor	Just bolt up motor, attach wires. and go No alignment issues	1.0 hours
Install 220V branch circuit 100'	Just hang	4.0 hours
Replace breaker panel	28 circuits, standard mounting, no problems	8.0 hours

Slotting Table

Job slotting is an effective work measurement technique because it is easier and more accurate to compare jobs than it is to estimate them from scratch. Slotting uses well-known jobs of various durations as standards or slots. Each of the standard jobs has a well thought through scope and planners and technicians have agreed upon the duration.

The job you want to estimate is compared with the standard jobs in the slotting table. The job is then moved up in the slotting table until the planner determines that it is longer than one slot and shorter than the slot below it. The estimated duration standard for the target job is set as halfway between the two slots.

Historical Averages

The labor-hours charged to previous work or individual jobs are recorded and accumulated. They are averaged after elimination of the high and low readings. The resultant averages reflect the size and condition of the facility, condition of the equipment, skill level of the maintenance work force, and the current state of job preparation and materials support. Because the work order system is the source of historical averages, it is often difficult to obtain reliable data on which to calculate the averages. In too many CMMS implementations, work content is charged to whichever work order is handy, not necessarily the right one. Another disadvantage is that the standards include all the lost time typical of that plant.

Universal Maintenance Standards

Predetermined Motion Times, Time Study, and Standard Data, evolved into Universal Maintenance Standards (UMS) and these have now faded from common usage. Although they are the most accurate method, such standards are too time consuming and expensive to set up as well as maintain. Each job is studied by an industrial engineer and analyzed for best practices. This technique should be considered only if you have to change 1000s of filters, rebuild tens of thousands of valves or some production-like job. Consequently, the approach is not generally recommended.

Flat rate manuals

Some industries (automotive, mobile equipment, HVAC) have repair times, techniques, and steps that are well-documented. When you buy a truck, for an extra $100 or so you can get a manual of repair steps, times, and other useful information. Ask the manufacturers of the equipment or components (trucks, construction equipment, seals, bearings, etc. for such information).

These times tend to be very well thought through but they often assume a level of tooling and work conditions that you might not be able to duplicate. The OEMs use the flat rate manuals to reimburse dealers for warranty repairs. You can be sure the times are quick (maybe too quick if you have older, dirty, or rusty equipment). Some dealers pay their mechanics based on these rates multiplied by their pay rate for all jobs they complete that week.

Engineered Performance Standards

Engineered Performance Standards is a special flat rate manual that was compiled by the US Navy in the 1970s. The Navy employed an army of industrial engineers to time study all maintenance work that took place on navy bases. The result was a library of manuals with accurate estimates for work such as painting, carpentry, electrical work, even standards for pier building! These manuals are available from the Superintendent of Documents.

Element 2 of a planned work environment: The Maintenance program

How many hours do you really have available for scheduled work? When we say 'really' we mean on the shop floor in reality (not in a manager's dream). This question is critical because if the answer is none then you cannot run a maintenance schedule. Usually the answer is a lot less than you would like.

The maintenance program outlines hours available for each category of work (the three crews mentioned earlier) each week. You might choose to run a maintenance program calculation annually for long term planning purposes. The work program is run weekly by the scheduler (or planner/scheduler) for serious use. There are four factors in the maintenance program.

1. The first calculation is total gross hours by craft. This is just an arithmetic calculation of gross hours, multiplying the hours worked per week per person by the number of people.

2. The second calculation is the total net hours by craft. Take the gross hours and delete vacations, holidays, meetings, personal time, training, a percentage for absenteeism, and a percentage for late starts and early quits. In short you delete any time where any tradesperson will not be available for work. Then add back any authorized overtime and any on-site contractor hours available for scheduling.

3. Deduct any PM or PdM activity from the net hours. PM comes out of the total hours first. These hours are scheduled and are not available for backlog relief because they are the workload of the PM crew.

4. Deduct a reasonable factor for emergency work. Some plants use 75% of last year's emergency hours for the week (assuming improvements in the effectiveness of the PM effort). We recommend an emergency crew be used for this workload. Some places use a small emergency strike force. These people are not on the schedule or are assigned to interruptible jobs.

The hours that are left are available for backlog relief. The totals are realistic because most of the traditional problem areas have already been handled.

Element 3 of a planned work environment: Individual maintenance job planning is an ongoing process starting with PM activity, then repetitive corrective jobs and finally all the rest of the jobs that are in the total backlog.

Advanced planning is a proactive skill. On larger jobs you can save 3-5 hours of execution time for every hour of advanced planning. Normally the pressure is on doing the job, not planning the job. Few users scream at you to plan repair jobs. This section shows the steps for successful maintenance job planning. The plan consists of a detailed list of the work to be done with all resources identified and an evaluation of each of the five elements of a successful maintenance job. Deciding on the scope of the work is the first part of the planning process. The goal of the plan is to enumerate all the resources needed for a job so as to eliminate the avoidable collisions between these resources.

The preliminary planning (in Element 1) is done simply to find out the estimated hours for the various jobs that are in the backlog. Once the preliminary planning is completed, the planner can proceed to the work program step that determines the hours available for backlog relief. When the hours are known the planner and his/her production counterpart can choose jobs to be scheduled for the next period.

After the likely jobs have been identified, more formal planning can be done for just those jobs. If this strategy is followed for several months, repeat jobs will begin to show up. The detailed plan is already completed for these repeat jobs. After 12-16 months, 90% of the backlog will have detailed job plans. As mentioned, if planners have some time before scheduling is initiated they are encouraged to plan all PM activity and common corrective jobs.

Answer these questions before planning an individual job (or as you plan the job):

✓ The first issue of maintenance planning might seem to be trivial but actually gets to the core of controlling maintenance activity, which is simply, what is the scope of the work?

✓ Is engineering required? If engineering is required, drawings and specifications must be ordered.

✓ What hazards exist? The work site must be examined very carefully for hazards. Some hazards might include working at heights, hot work, bad weather, working near energized equipment, etc.

✓ The planning function also assigns the degree of importance of the job by answering the question "what is the priority of this job?" Some automated systems assign priority using a mechanical means and in some instances the requestor recommends a priority but the planner must find out if anything will modify the mechanical priority.

✓ What are the work steps? Qualified planners will visualize the steps that must be taken to accomplish the job. They will specify mechanic(s), technicians, and helpers: define what skills are needed, how much craft coordination, time per step, crew size, and what contractor(s) will be needed. Each step must be visualized, along with time, labor, and other resources.

✓ What tools and equipment are to be used? What special tools for each job step (include equipment such as cranes), where to procure, how to ensure availability, name of vendor for rental or purchase.

✓ What materials/parts/supplies will be needed, and of course, how many. When the bill of material has been prepared, the planner establishes availability of the parts and if they are carried in stock. If not carried in-house, who is the preferred vendor and what is the lead-time. In some plans, the requisition for the non-stock part is written (but not turned in until the go-ahead is given).

✓ Depending on when planning is being done (for an immediate job verses planning to complete the planner's file) there might be an implied go-ahead so that purchase requisitions would be issued by the planner at the time of planning.

✓ How long will the job take? By adding up the times for each step, the planner establishes the likely elapsed time. Explicit in the job step information are crew sizes for each step.

✓ What is the probability of something going wrong? The job could end up being significantly larger than first thought or there might be non-obvious hazards. Create back-up plans if the job is critical or high priority, if the scope of work isn't adequate, if the job doubles or triples in size, or new hazards are uncovered, etc.

✓ What drawings are needed? Collect any drawings, diagrams, sketches, and wiring diagrams.

✓ Availability of the unit to be serviced: When is the best time to do the work? Look for the best time in the business cycle for a convenient time to repair an item. Determine the effect of this repair on related units. Will a partial or complete shutdown be needed?

✓ What authorizations/permits/statutory permissions are needed? Possibilities are hot permit, open line permit, tank entry, lock-out/tag-out, EPA involvement, Safety Dept

After all the questions are answered we end up with a planned job package. The list that follows includes many items that are only appropriate for the largest maintenance jobs.

Contents of a planned job package (select items needed for individual job)

(adapted from Managing Maintenance Shutdowns and Outages published by Industrial Press)

Planned job package table of contents (for larger jobs), or cover sheet

❑ Detailed Work Order (or work orders if the job is large) spelling out the work to be done. In some organizations, the planner or operations person physically hangs tags where work is to be performed. For smaller jobs the work order is the plan and is complete in itself.

❑ Job Planning Sheet with Sequenced Tasks detailed by craft and skill level. Contractor(s) as well as in-house resources must be included.

❑ Lockout and tag-out, confined space entry, or other safety instructions if they are not included on the work order.

❑ JSA (Job Safety Analysis) describing hazards, safety requirements and PPE requirements. Keep in mind that the best approach to safety is to eliminate the hazard. The second best is to build safety into the job plan from the beginning.

❑ Requirements for decontamination, de-watering, cooling, and removal of product.

❑ Details of all pre-fabrications. Estimates of the lead-time for the prefab work.

❑ Duration and labor-hour estimates for each Task. When beneficial, (particularly for longer, more complex, or critical jobs) a Gantt or CPM chart should be prepared to convey trade sequencing and simultaneous tasks to the assigned crew(s).

❑ A Bill of Materials including availability, commitment and staging location (for large spares). The listing should distinguish between authorized stock items, direct purchases, indirect purchases (contractor will buy), in-house fabrication, and outside fabrication.

❑ A list of all specialized tools, machines and heavy equipment. This list should include everything from cranes, to line freezing rigs, to scaffolding. It also includes power tools that are common but not normally carried by the trades such as portable band saws, threading machines, or impact hammers. The list would also include standard tools that are needed in large quantities.

❑ All clearances and all required permits (both governmental and company) completed to the point of safe feasibility. These documents might include permits for flames, confined space entry, open line, valve close/open, discharge, building, etc. Of course, the final lock outs must be made by the responsible mechanic and equipment operator

❑ Outside specifications (and standards) that apply to the job such as ANSI, ASME, etc.

❑ Other reference documents that the assigned crew is likely to need, such as prints, sketches, photos, specifications, sizes, and tolerances.

❑ O & M (Operations and Maintenance) and other manuals or sections of manuals as required.

❑ A Site Set-Down Plan (for major tear downs).

❑ Sign off sheets for job completion from worker/contractor, data entry, safety, and others as appropriate.

Larger jobs might include most or all of the items listed above. Small jobs might just have a work order with a list of parts and tools.

Planner's job

Being a Planner is a very interesting job. You have the intrinsic interest of what is happening on the shop floor with the satisfaction of planning and laying out efficient ways to work. Most planners also schedule, and the scheduling part of the planner's job is discussed in the next section.

Typical Activities of a Planner

- ❑ Maintain packages on all jobs planned but not started.
- ❑ Inspect jobs and discuss the scope of the work and its priority with the customer.
- ❑ Conduct JSA (Job Safety Analysis) if you are trained and expected to do that
- ❑ Look into history files to see if this or a similar job was done previously.
- ❑ Determine the scope of the job. Make sketches or take photographs if necessary.
- ❑ Break job into smaller projects or job steps.
- ❑ Estimate those job steps
- ❑ Prepare Work Order. Get authorization if it exceeds your authority.
- ❑ Write-out work plan (steps to complete project)
- ❑ Locate all materials in stores
- ❑ Determine lead times and vendors for non-stock materials.
- ❑ List special tools and equipment needed for the job.
- ❑ If asked, plan longer term requirements such as yearly PM requirements by month.
- ❑ Verify that the job was done according to the plan. When a job deviates, learn why. If necessary update plan. Accept feedback from the field about the job plans

The Planner (or the planning function) is a key to the effective daily operation of the maintenance department. Competent planning requires a wide range of skills. Some resources showed the average ratio of planners to mechanics in various crafts in heavy industry as follows.

Electricians	1	to	20
Welders	1	to	11
Machinists	1	to	14
Riggers	1	to	24
Pipe fitters	1	to	17
Laborers	1	to	30

Ratio or Planners to Craft People

Qualifications needed to be a planner

1. Ability to think of jobs as having both mechanical aspects and abstract (time, tools, space requirement) aspects.
2. Ability to express themselves both verbally and in writing.
3. Gets along with others.
4. Can work with different types of people from all levels in the organization.
5. Can represent the organization's interests in discussions with outside firms.
6. Has some knowledge and skills in one or more crafts.
7. Is interested in the other crafts that they are not familiar with.
8. Has the respect of craft workers.
9. Has a positive attitude toward company, supervisors, and managers.
10. Has the ability to plan work and foresee problems (doesn't like surprises).
11. Understands or can be taught job planning and scheduling.
12. Computer literacy
13. Some training in budgeting
14. Understands the issues of the user department

Element 4 of a planned work environment: Coordination with production

One of the most important (but smaller in time commitment) chores is to meet with your production counterparts and agree on which jobs from backlog will be worked upon next week (or next schedule period). You come to the coordination meeting with a print out of the ready backlog in priority order (or on a handy screen). You also bring a print out of all (interruptible) PM and PdM work orders scheduled for that area. Non-interruptible PMs do not impact production (can be done with the equipment running) so they do not have to be discussed at any length.

You also bring an up to date copy of the current week's schedule. Any jobs that are going to be carried over from the current week generally get top billing on the next work list.

At the coordination meeting you and your coordination partner, the production planner, agree on each job that will go on the work list for the next week. The PMs are automatically on the work list but do have to be given slots by the production planner. First you add carry over jobs to the work list. Then you go down the backlog list, adding jobs to next week's work list until you run out of hours (the hours you calculated that were available for backlog relief). If specific times and dates are important, they are noted on the work list.

This work list will be signed by both parties. The production planner's signature means that he or she will do their utmost to insure that the equipment will be available for maintenance work at the agreed upon time and date. The maintenance planner's signature means that the appropriately skilled maintenance person with the tools, equipment, and parts will be available to work on the agreed unit at that time also. The signatures also mean that, if the situation changes, the planners will communicate the new conditions to each other.

The coordination meeting has another function. Last week's schedule compliance is reviewed. If the compliance is low then each job missed should be discussed with an eye toward solving the underlying problem (not to affixing blame).

Where there are on-going problems with broken commitments, attendance at the coordination meeting should be escalated to each coordination partner's boss. The bosses should sign and accept the work list. If problems with commitment continue, the plant manager should host the coordination meeting. In any event the plant manager, maintenance manager, and operations or production manager, should attend part or all of the coordinating meeting.

Element 5 of a planned work environment: Scheduling the work for the next week

The work list is the first step in scheduling. In fact in many plants a coherent work list for next week would be a tremendous improvement. Keep in mind the work list is in priority order and has been adjusted for available hours (in the work plan).

Why Scheduling improves maintenance department performance

All the elements of the maintenance job from the job plan are brought together in precise timing

The schedule allows the supervisor to concentrate on managing the issues of each job. To bring the whole job under control you control the parts. The schedule contains timings of all jobs so that the responsible supervisor can manage at the time and at the level where they can have a useful impact.

The benefit is that the supervisor has the information to monitor the whole operation; repair by repair, mechanic by mechanic, and day by day. Monitoring means keeping ahead of jobs that are going to start, and checking on jobs in process.

For example, supervision knows from looking at the schedule that a large repair is falling behind and missing milestones. With this knowledge the supervisor can intervene so that the situation can be corrected while supervision can still have an impact on that job. Waiting until a monthly meeting to find out that a job went overtime is too late.

Another way to stress the importance of scheduling is to say that the schedule organizes the attention of supervision and applies the attention where there is a problem. It provides a kind of 'management by exception'. The supervisor is focused on the problem jobs where he/she can apply their problem solving skills to a real problem at hand. In short the schedule helps a medium quality supervisor become a good one.

Frequently the reason that a schedule miss is occurring is related to the mechanic not having a critical piece of information such as special tools, techniques, general experience, or a possible material substitution. The supervisor, having been informed, can intervene and get the job back on schedule. The information is passed back to the planner so that the information is available on the next go around.

People work better when they are given a reasonable amount of work to do each day. Work is quantified in the job plan. A reasonable amount of work is expected from each worker each day. Workers are freed from a hurry atmosphere one day and a kill time atmosphere the next. Management's expectation as to how much work is a reasonable day's work is given to the mechanic in advance. Intermediate goals are identified and checked by the supervisor.

There is a psychological reason why scheduling works and why it improves the quality of life in the organizations that use it in an enlightened way. These reasons are rooted in psychology and one reason at least goes back to the early pioneering studies at the Chicago Hawthorn plant of Western Electric The study found (to compress a huge body of work into a few sentences) that workers at Western Electric responded with increased productivity to attention from management. The classic study concerned the effect of lighting on productivity. When lighting was increased, productivity went up. When lighting was decreased, to everyone's surprise, productivity also went up. After many experiments the attention factor was found to be critical. Maintenance schedules provide structured attention to the maintenance workforce.

An inevitable by-product of this approach is the uncovering of many hidden problems with the business system that runs maintenance and production. Planning highlights areas where mechanics cannot do their job due to a problem outside their control and other problems with the old way of doing business. These previously hidden (or unpublicized) problems suddenly come into the foreground (how many are true in your factory):

Contributors to Schedule Misses

✓ Failure of the equipment to be available when promised.

✓ Mechanics being pulled off to work on non-productive activities. This is a common problem where the maintenance department is also used in personal service to the plant higher-ups. Mechanics running to the airport, picking up contracts, or setting up for picnics, cannot adhere to a work plan, or stay on schedule (unless the activity was scheduled with the maintenance planner the prior week!).

✓ The planning package is horrendous. The planning package does not reflect reality. Parts are wrong, job steps are outdated or wrong, lock-outs and new regulations are not included.

✓ The stock room regularly contributes to schedule miss conditions. Items shown on the inventory are not on the shelf. Quantities are wrong. Stock-outs are frequent on regularly used items. Excessive time is spent waiting for clerks. Parts cannot be found or take excessive time to find.

✓ Failure to put equipment back into service when promised. This failure builds the attitude in production that they never want to give up a machine because they never get it back when promised. This attitude stands in the way of good communications and contributes to the problem of an inability to get control of production equipment when scheduled.

✓ Lack of cross training causes clashes of resources. You might have enough maintenance people but you are chronically short of specific skill sets such as computer or instrumentation skills. The cause is lack of flexibility because of single-skilled tradespeople. There are some good arguments for craft shops. We have to face the fact that a craft shop is less productive than a specialized maintenance department, and often adds an additional potential collision (enough people but not enough of a critical craft) to reduce productivity.

How to set up a schedule

Set-up ahead of time:

All crews, teams or individuals are assigned unique names/codes. Build a spread sheet template with the crews or people down the rows and the days of the week across the columns. Each day should have 12 columns (for a 24 hour operation) or 4 columns for a single shift. Each column is 2 hours. The first two columns should be work order number and short description (a third column for priority is also helpful).

Review all jobs on the work list from the coordinate meeting. Load the schedule spread sheet starting with the jobs where a fixed date and time was agreed to. After loading all the fixed jobs fill in the rest of the time, starting with the highest priority jobs that remain.

When loading the jobs into the spread sheet you chose a crew by filling in the job in their row. Remember the work plan allowed time for emergencies.

You have available a percentage (equal to your historical emergency hours) of your crew's time unscheduled.

If a high priority job cannot be scheduled because of inadequate specific skill sets, or scarce resources (two jobs needing the crane) try to juggle the jobs to different times in the week. Investigate other solutions (contractor, overtime, equipment rental). If there is no obvious solution, inform your coordination partner of the problem.

The schedule should be updated every day to visually reflect the current status of all jobs. Big problems should be communicated to your coordination partner. Twice each shift, checks are to be coordinated with shift changes and meal breaks, and all jobs are to be evaluated for status relative to the schedule. Jobs the schedule show complete should be complete. Milestones should be met on longer jobs. Shorter jobs that span a check are to be informally reviewed.

Jobs of the Scheduler

Job tracking:

- ✓ Build the schedule from the approved work list
- ✓ Assemble individual planned job packages to be given to the supervisor with the schedule
- ✓ Insure safety instructions are in the hands of the craftspeople or their supervisors
- ✓ Maintain a tickler file on all projects that have been started.
- ✓ Notify and consult with customers and your coordination partner about any pending interruptions or disruptions
- ✓ Get information about status of all repairs on a regular basis. Update physical schedule at least daily
- ✓ Be alert for continuing schedule miss conditions. Detect when a job runs into trouble before it misses a milestone.
- ✓ Attend weekly coordination meetings with planner to discuss work progress, progress on projects, and updates on materials.
- ✓ Clean-up paperwork at the end of the jobs.
- ✓ Complete daily, add to weekly reports for higher management.

Material/parts/supplies tracking:

- ✓ Follow-up on availability and delivery of parts for planned work orders.
- ✓ Call vendors when parts are late, and to get parts that might slow down or stop a scheduled job.
- ✓ Get the purchasing agent involved when a vendor is being unreasonable or is suspected of lying.

✓ Keep a check on the most-used and critical items in stores. If appropriate, place parts in a protected area.

✓ Work on assuring more accurate usage information on parts. Update the work plan with actual part usage, or give the information to the planner

✓ Update the plan (or inform the planne) of usuage os supplies, rigging hardware and open stock items used.

Maintenance Work Order

Many methods are used to collect information about repairs. They range from manually generated paper-based work orders to computer-generated work orders, and from work downloaded to a PDA (Personal Digital Assistant like a Palm Pilot) to wireless laptop work stations. Use of outside technology such as bar code scanning, RFID (Radio Frequency ID) chips has simplified, quickened, and made data collection more accurate. Collecting the work order information data has been in the past the most expensive aspect of having a maintenance management system.

Primary Use: Communication with mechanic. Work Orders give the mechanic written instructions that include the asset, its location, a description of the problem, who to see, the contact's name and phone number, authorization to proceed, and fixed information about the asset. The Work Order also gives all the information from the planning function including the bill of material for the repair, tools, equipment, and engineering data.

You need a work-order or a write-up sheet (showing what work was done, what parts were used, and what outside or rented resources were needed), because after communicating instruction to the mechanic, the write-up is the basis of all analysis. Every hour worked by an employee or a contractor should be accounted for (after all, you paid for them). The work orders form your unique history. Having them easily available for research or review (by computer or even manually) pays dividends in root cause analysis and continuous improvement programs as well as all kinds of decision making based on data.

The work order can also be a tremendous benefit in the following areas:

1. Cost collection.

All the labor and parts are recorded on the work order. Charging all time and materials to the appropriate job creates a more accurate view of the real profitability of a process. In addition to parts and labor, outside services, equipment rentals, and rebuilding should all be charged to the appropriate work order. Cost is collected by work order number and then can be allocated by asset number, line, process, and department, or by plant (or by some combination). The work order can also be used as the basis of a charge-back system. In a charge-back system the work order is like an outside vendor invoice and is charged against the originating department's budget. Charge-back is an excellent way to highlight user departments that use an inordinate amount of maintenance resources (and ones that use little resources).

2. Proof of safety and proof of compliance with codes and statutes. Operating licenses require certain checks and rebuilds to be made on a periodic basis. The work order becomes proof that the event or inspection actually happened. The scheduling part of the work order system (usually under the PM subsystem) helps manage future inspections or work required for the license.

When there is an accident or some other catastrophe, officials will want to see your maintenance records for the area. While having records (and having done any indicated work in a timely manner) will not remove the pain of a catastrophe, being able to prove that you did what you could, will make you feel somewhat better. It will show the authorities and your management that you were doing your job.

There is a dark side to using the work order system for proof of compliance with safety or environmental regulations. What happens if you chose to defer work on deterioration that your inspector uncovers? Having a work order system clearly showing inspections were ignored and work was left undone is a serious problem. Casual negligence becomes gross negligence or worse.

Whenever there is an incident, the maintenance records provide excellent information for the investigation. In a few instances the accident could be the result of inadvertent consequences from the maintenance procedures themselves. The best example of this was the failure of the engine pylons on the DC10 aircraft. Maintenance crews using lift trucks to remove and replace the engines. were lifting the wing (when they lifted the engine too high), and caused cracks in the mounting brackets (pylons). One of the cracked pylon brackets failed and caused the crash of an American Airlines' jet out of Chicago.

Properly executed PM records can significantly reduce your share of the liability if you are involved in litigation. The truth is that a plant with a well-documented maintenance effort will both be safer to work in and is more likely to be in compliance with codes and statutes.

3. Job, project, and mechanic management, scheduling, and control document.

The work orders show the progress of a large job and facilitate control. On a large job you can write all the sub-tasks on the work orders and use the forms to manage the job. Each work order can be an activity in a large job or a shutdown. The information about duration, precedence, and resources can be entered into a Project Management System (PMS). Companies use the PMS to create a Gantt chart or network diagram. As work orders are completed, the activities can be updated, which will update the charts. The up-to-date charts are excellent management tools.

4. Lets the user know what is happening on their job.

Work orders aid communication with the user. Sending a copy of the document after it enters backlog shows the user you are working on their problem, improves user relations, and gives them a work order number to track. Some systems handle this distribution by E-mail and allow the originator to track the service requested to a work order, and the work order through its stages. The system will notify users when the work is completed. On manual systems we recommend that you send another copy or an E-mail message to the user when the job is complete.

Follow the example of a major pharmaceutical company and put a quality/satisfaction survey on the back of the user's copy of the completed work order. When the work order is folded, the survey is on the outside. Questions include; were you satisfied; was the job done in a professional manner; was the mechanic on time with any appointments, any suggestions.

5. Provides history for various assets and their components.

The history derived from the work orders tells you such things as: which types of compressors last longer, which valves are easiest to rebuild, and many other useful bits of information. It only takes a few years before you have enough history to guide decisions. An interesting aspect is component life analysis expressed as MTBF (Mean Time Between Failures) for major components. In some systems, if you have a class of similar equipment in similar service, you can aggregate the data for greater accuracy. History can also be used to compare different conditions (such as different PM intervals, or different types or frequencies of lubrication).

6. The work orders might jog your institutional memory.

The work order history also is useful when you have not done a repair for an extended period and the work order from an earlier repair is available. In some instances (more often recently) no one in the plant personally remembers the job. The Work Order could be helpful in locating vendors, finding part numbers, preparing job plans, and for research into techniques. The key is to have preserved any notes and small drawings that find their way into the file and to have them attached to the bottom or back of the paper document. New computer systems allow images to be scanned and attached to a completed work order.

7. Authorization for parts and special tools.

The work order becomes authorization for all resources consumed on the job. The order is authorization to the supervisor to use labor; and authorization to the storeroom to release the part. The order could also be used by purchasing as the initiator of a material requisition, if the item is not in stock. The work order can also be the document that allows a manager to hire contractors or rent heavy equipment. If contractors are used, the order allows them to buy materials against your company account and secure special tools.

8. The Work Order can become a training document for less skilled workers.

By reviewing what was done for each type of complaint the new mechanic can see what kind of problems someone had to fix, and what parts were used.

Summaries of work orders can be used to design a training course by pinpointing the most likely repairs for training. These same summaries can be used to send new maintenance workers to school, or to monitor the progress of a new worker (when she completes rebuilds of each of 5 pumps then she can work on pumps on the line). If the work is formally planned, the work order will have job steps that might also be useful for training.

9. Provides data for root failure analysis.

When a recurrent problem comes up, the work orders will show you how often, how expensive, what parts were used, and how long the previous repair took. This information helps identify the financial exposure (first step). The dates, notes, work accomplished, and parts used, provide data for root cause analysis. Without a work order you have to rely on the memory and conversations of the mechanics that have worked on the unit (who may be tough to get into one place since they might be assigned to different shifts).

10. Insurance recovery

If you have a fire or other casualty claim, the insurance company can use the write-up to determine the scope of work and the hours and resources used, to determine the amount to pay you. If you claim that an insurable event caused disruption to maintenance activity then an analysis of the work orders might provide the proof needed by the insurance company.

11. Warranty recovery

Many items that you buy have warranties . In factories across the globe, millions of dollars of warranties are unclaimed. The work order can sometimes be used to document the problem for warranty recovery. In some instances, a large group of premature failures could initiate warranty recovery for all items in that batch. Some warranties include labor and some only include materials.

In a large fleet operation (which as an industry has always been careful about warranty claims and collections), a new design of Refer- refrigerator unit (used to cool a trailer transporting frozen food) had significant maintenance problems. The manufacturer disputed the amount and depth of the problem. After the company analyzed their CMMS records they realized that eventually all the units would break and could cause extreme damage to loads in transit. They decided to apply for warranty relief for all the units they had bought, even though only a few had broken up to that point. Armed with detailed work orders, they met with the refer manufacturer. After seeing the scope of the problem on paper, the manufacturer not only granted the claim for all the units, he supplied extra refer units so that the fleet could swap the bad units out and repair them on the bench. Once alerted to the problem by this fleet, and after an investigation into other fleets, the manufacturer issued a service bulletin and repair kit.

12. Feedback to planners

The work order tells the planner what actually happened and how long it took. In some areas the work order can capture other resources used (which are added into the master plan for that job). This feedback can improve the effectiveness of the planning function. To secure the information, many facilities that have planners close out work orders and enter the data into the computer.

13. Source document to reconstruct what really happened

We need to get to the bottom of every expensive or disruptive failure. A properly filled out work order will facilitate this analysis.

14. Source of data about hidden demands

The maintenance department is frequently recruited into playing roles such as driver, pick-up person, furniture mover, personal servant, boat polisher, picnic set-up/clean-up crew, security crew, and so on. These are special jobs that need to be tracked. A southern manufacturer that hired a limousine company to pick up visitors at the airport rather than use their maintenance mechanic, saved significant money. The savings were in time lost, parking, and disruption to on-going jobs, auto costs, and insurance. Most of the passengers preferred to be picked up by professionals (not to mention the preference for a spiffy town car over a typical mechanic's dirty pick-up truck).

15. Evidence in court

The maintenance information system provides data for court hearings on liability claims, injury claims, breach of contract claims, etc.

Design of work orders: If you design your own document be sure to include:

1. Wide lines or double spaced printing so that the mechanic can write in easily.
2. Size the form to fit on a clipboard if you expect the mechanic to write on it (or buy special clip boards).
3. Have your organization's name printed on top to be more professional.
4. Use check off boxes where possible to reduce writing and improve completeness.
5. Consider using bar codes to speed data entry. Bar codes also increase accuracy.
6. Consider the paperless Work Order systems where everything is typed or scanned into terminals around the shop. Alternatively, use hand held PDAs with synching during breaks and lunch, or wireless equipped laptops that are radio linked to the network.
7. Provide room for comments and sketches and be sure these diagrams get into the history for the asset.

Quality control for Work Order Systems

It cannot be stressed too strongly that all the benefits from a CMMS depend on the integrity of the data entered into it. Without complete and accurate data, decisions will be affected by whatever data didn't make it into the database.

In an iron ore mine with a large maintenance department there were two Data Integrity Officers. These people reviewed almost all the work orders coming back in from the field before they were re-entered into the system to close out the work. If anything was missed, and if any data seemed wrong, they marched the offending document back to the accountable supervisor for correction. They also managed the task of accounting for all the work done. The concept was great, and it was helped by the choice of people for the job. Management picked people who everyone was afraid to hear from as the Data Integrity Officers!

When you analyze work orders, note that there are two parts (at least) to every work order. Each part should be studied because problems in the different parts will have different solutions. The two parts are the header and the body. The header is filled out by the person requesting service, and the body is filled out by the person who will be doing the work. Although the header can be filled out by operations (for a breakdown), or a PM mechanic (for corrective work), the body is always filled out by the mechanic doing the work (or sometimes by their supervisor).

There are two issues: completeness and accuracy

Completeness: Are all the fields filled out?

Accuracy: Is the data in each field accurate?

An easy way to tell if your organization needs intervention is to review a random stack of work orders or requests for service (or what ever it is called in your system), before they are edited or modified by anyone in maintenance. Periodically review about 100 randomly-chosen work orders.

Work Request Header Audit

During this review create a check sheet adapted to your system like the one below:

Number of Work Orders Reviewed (be sure to choose them randomly): ____ = Total

Date Reviewed________________________ Reviewer______________________________

Review this field	**For this information**	**Checkmark if the work order fails**	**# Fails (F)**	**% F/Total**
Work requested	Specific, observable, actionable			
Asset #	Correct, readable, same name as in CMMS?			
Reason for repair	Is the category correct?			
Requestor information	Complete and accurate including phone, E-mail for questions?			
Priority	Does the priority make sense giventhe job requested?			
Timing (when needed)	Is the date needed reasonable, given the request?			
Location	Is the location where the work to be done clear?			
Proper account to charge if needed	Is the account correct, is it filled in if necessary?			
Other________				

Depending on what you find, if problems exist, consider training. If only a few people are involved, informal training might suffice. If there are a number of people entering work requests then formal training is in order. What about conducting a Work Request training class? A formal class built into orientation is important if the people that are expected to fill out work requests are turning over quickly.

Training course for filling out Work Order (work request or notification) header

Do you have a training course for how to fill out work orders, maintenance notifications, or service requests? Many organizations have "fill out the form" training while glossing over or skipping entirely the essential issues of accurately describing the work to be accomplished and the fields that facilitate analysis.

Two groups need this training. The biggest is the group of operators and other maintenance users. These people are usually writing work orders or work requests based on what happened last time. They also usually don't have significant formal training in maintenance (although they might be very smart in these areas).

The second group consists of the maintenance workers who perform PM duties and find themselves writing up corrective maintenance requests. These people would be expected to have significant maintenance background. Although training is essential for PM workers, all maintenance department employees should have significant expertise in the work order entry part of the system. They should be shop floor experts and able to help the production and other user groups.

Sample curriculum for a work order training class

Additional information for your class on filling out work orders can be found in the next section, which reviews each field of the work order and each type of work order.

Work requested This is the field that usually needs the most training. Operators tend to put their conclusion down (pump broke) or their solution (replace pump). While this might be ok some of the time a more useful approach is for the operator put the specific observable phenomenon such as 0 gauge pressure on outbound side of pump 3S on the Stoker line. They might add in a comment (I think the pump is broke).

Asset # In some instances the shop floor operators have different names for the machines than do the maintenance department files. It is important that everyone use the same language and referral names. If necessary, paint the name on the side of the machine or (if possible) put a cross reference chart into the CMMS.

Reason for repair The reason for the repair is an important field for future analysis. Any kind of improvement project should be distinguished from regular maintenance. Any maintenance done as the result of an inspection activity should be coded differently from jobs found by breakdown.

Requestor information Can you reach the requestor with the information on the work request? If it was E-mailed then the reply address should be good.

Priority There should be an established, well-understood, priority system. The request should conform to this system. If your system does not allow the requestor to set, even preliminary priority, then skip this field.

Timing (when needed) As with priority, timing should be associated with each type of work. Does the timing requested conform to this scheme? If not, is there a compelling reason for it to not confirm to the scheme?

Location Is the location for the service accurately described? In some systems, an accurate asset number will automatically bring up an accurate location. If so, then skip this field unless it is a problem for some other reason.

Proper account to charge if needed Are any needed codes correct? Some jobs require accounting codes for proper allocation of maintenance and repair dollars. Some firms charge back maintenance to an originating department. All projects should have accounting codes if they are not maintenance activity. In some factories, a work request with HOT or RUSH would charge overtime to a user account.

As with all kinds of training, design your classes to teach how to fill out work requests, conduct practice sessions, discuss the consequences of badly filled out documents (show examples), drill, follow up with auditing and coaching. Be sure to give plenty of practice and display some good work orders already in the system.

One important part of the training is to show users how good work requests result in better service for them and better problem solving in the future. Conversely show how badly filled out work requests make good customer service difficult. The next section discusses auditing the body of the work order and the training curriculum.

Four types of work order forms serve different purposes

1. Maintenance Write-up Form (MWU)

The PM inspector or surveyor fills out the maintenance write-up form during the inspection. A maintenance office worker can also fill it out for a called-in complaint. This type of work order is the most common. During an emergency, the documentation would be completed when the job is completed. Jobs occupying less than 30 minutes or 1 hour would be sent by radio to the mechanic and details entered on the maintenance log sheet (next form).

As described above, maintenance departments need to conduct periodic classes in correctly filling out this document. Everyone who is expected to fill out work orders should be trained to write a complete description of the problem in question and be informed which other fields need to be addressed. The better this form is filled out, the more likely the user will get what was wanted, and the easier and more accurate will be any subsequent analysis.

2. The Maintenance Log Sheet (MLS)

The maintenance person does many jobs that take only a few minutes. Initiating a full work order seems to be paperwork overkill. Blowing off the time is a problem too. Major problems are sometimes covered up by numerous short repairs or adjustments. Simple problems are never addressed because they never take enough time to be escalated to full review by the work management system. These little things can add up to significant dollars in a year. Not completing short repairs in a timely way is also the main reason for user dissatisfaction and subsequent maintenance manager hassle. The Maintenance Log Sheet (MLS) is a good compromise.

The MLS could be carried by craftspersons to record all the short repairs that they do in the course of their day. Included are informal quickie requests made directly by operations (maybe when the mechanic walks by), or calls from the dispatcher or supervisor to repair or adjust something without a work order number.

A similar form might be used by operations to record minor problems and fixes. This work would include TPM (Total Productive Maintenance) attempts to record intermittent work stoppages and what was done to repair the stoppage. These stoppages are usually corrected by short unrecorded adjustments and repairs, and this document is used to record such small problems. The form could also be carried by a regular vendor to record all the little things that they complete.

The log sheets are summarized and entered into the CMMS on a weekly or monthly basis. If there are only a few assets, the mechanics would carry one log sheet for each asset. These records can be summarized into a single work order for each asset. If there are many assets then the work should be recorded in chronological sequence.

3. The Standing Work Order (SWO) or Blanket Order

Thw Standing Work Order form can be used for jobs that are done routinely with known labor and materials. Examples would be a machine start-up every Monday morning, or the safety walk-through. A single Standing Work Order might be good for a week, a month or longer. These items are usually routine jobs that aid the user. The standing work order can be virtual, that is, just a work order kept open on the CMMS (if the system allows this). That way, losing the work order is harder and hours are booked when they are expended.

We want to capture the event, the time, and the materials, while spending as little clerical effort as possible. We can effect this economy because the work content of jobs that would qualify for standing order status is well known and understood.

Do not abuse this concept with unrelated work charged to the standing work order number. A power plant in the West had an entire department of 16 workers charge their time to 21 blanket orders (one for each pulverizer) each month. As a result, the work order system contained very little useful data.

4. The String Work Order (STWO)

Most work order systems are designed to permit time and materials to be charged to a single unit or machine. String PM's are the only work orders that are not unit based. The string is a group of like activities done on a string of assets that are strung together for PM purposes (like pearls on a necklace). A vibration route, or filter change route, are examples. The string PM form is usually used for simple tasks (that are not time-consuming) repeated on each asset.

The string PM is one of the hardest to account for in computerized maintenance systems (how do you divide the time). Most CMMS allow the input of only one asset number per work order. Some systems will allocate time to several assets evenly.

Note: A PM task list could be any of the three types of form. Most PM's use a standard write-up form that is closed out when the individual PM is completed. The form used might be a special one for PMs only but it will function like a standard write up. This specialized form for PM activity might have spaces for readings, check boxes, and places for specific notes. One advantage is the ability to add a task line item to record short repairs.

The standing work order form can be used for frequent PMs where the PM is done and the initials of the person and time are all that are required (you see such record forms at the airport for the hourly bathroom checks, or in operations for walk-arounds).

A string PM work order can be used for inspection routes and lube routes.

Examples of the types of Work Order follow

Explanation of All fields on the Work Order forms

Maintenance Write-up (MWU), Figure 16 used for most jobs

Maintenance Log Sheet (MLS), Figure 17 used for short jobs

String Work Order (STWO), Figure 18 used for simple jobs on many assets

Standing Work Order (SWO), Figure 19 used for repetitive jobs

All work orders are divided into three sections:

Header: Information known before the repair starts. This information is supplied by the originator or taken from the master files of the CMMS. Always have the mechanic verify the fixed data such as nameplate information, as the first step on the job.

Body: Information from the planner including materials needed, time needed, tasks, etc. This form is also used to record feedback from the mechanic of time and materials used.

Summary: CMMS or clerk calculate the summary information. In a manual system, the clerk extends the time, multiplies against the charge rate for the shop or for the worker, and adds up the dollars and hours. On a computer system, the system does the calculations and looks up the prices of materials.

Header of Work Order (information from requestor or from the CMMS master files is usually known before the repair begins):

Action Tip: The fixed files in the CMMS are entered at the time when the system was set-up. Subsequent equipment purchases and retirements are modified periodically. The information is usually 90-95% accurate. We want to improve the accuracy of the fixed information in the CMMS database. One way to improve the quality of the data is to have the mechanic correct any mis-

MAINTENANCE LOG SHEET FOR SHORT JOB

MLS#

DATE ____ TO ____

NAME:

LINE DATE	ASSET	LOCATION/ SYSTEM	USER NAME	WORK REQUESTED WORK COMPLETED	REAS/ PRITY	MATERIALS	IN OUT TOTAL
1							
2							
3							
4							
5							
6							
7							
8							
9							
10							

maximum job time 30 minutes, prepare write-up over 30 minutes

Figure 16

#MWU	LOCATION:	DATE:	TIME DOWN:
MAINTENANCE WRITE-UP		DOWNTIME? Y N	TIME BACK IN SERVICE
USER:	PHONE:		DOWN HOURS:

PRIORITY	100 DGR FIRE SFTY	80 STOP PROD	70 SAFETY OR CODE VIOLAT.	60 PM / DAMAG.	50 EFFIC IMPROV.	40 COMFOR	30 COSMETIC / CHG USE	OTHER PRIO	SPECIAL LOCK-OUT

REASON FOR WRITE-UP	SCHEDULED WORK											UNSCHEDULED						PERMIT REQUIRED
	PM	CM	UM-F	R	RM-M	I	E	U	OTHR	CL	GN	UM	PS	DR	DU	MU	OB	CONFINED SPACE
																		OTHER

SYSTEM:	REQUESTED BY:
	DATE REQUIRED:
	CHARGE-BACK ACCOUNT:

DESCRIPTION OF WORK REQUESTED:

SKILL LEVEL	UNSKILLED	MAINTEANCE PERSON	LICENSED TRADES	ENGINEER OR OTHER	CONTRACTOR:

TIME IN	TIME OUT	HRS	DESCRIPTION OF WORK	TOTAL PRICE	PARTS & MATERIALS: DESCRIPTION/PART NO.	QUAN
TOTALS			* CHGRT ( ) = +		= TOTAL THIS W/O $	

WHAT WAS FOUND: NOTES FROM MECHANIC

DATE COMPLETED: INSPECTED BY:

Figure 17

#MWU	BUILDING, LOCATION OR ASSET	WHEN DONE CHECK OFF	DOWN TIME	COMMENTS
STRING PM **WORK ORDER**	1.			
	2.			
DATE	3.			
	4.			
COMMENTS:	5.			
	6.			
	7.			
	8.			
	9.			
	10.			

PRIORITY:	70 SAFETY OR CODE VIOLATION	60 PM / DAMAGE	50 EFFIC. OR IMPROV	OTHER	DOWNTIME REQUIR Y N HOURS:	SPECIAL REQUIR

REASON FOR WRITE-UP:	PM	UMR	CL	GN	OTHER	DATE REQUESTED:	LOCKOUT PERMIT CONFINED ENTRY OTHER

CHARGE BACK TO ACCOUNT	REQUESTED BY:				
	SKILL LEVEL	UNSKILLED	MAINTENANCE PERSON	LICENSED TRADES	ENGINEER OR OTHER

DESCRIPTION OF STRING:

TASK:	MATERIALS:
HOURS: CHARGE RATE:$	MATERIALS $:

TOTAL THIS W/O: $

Figure 18

#MWU	**STANDING WORK ORDER**	DATE OPENED	DATE CLOSED
	USER: PHONE:	LOCATION:	SPECIAL LOCK-OUT PERMIT REQUIRED CONFINED SPACE OTHER

PRIORITY:	70 SAFETY OR CODE VIOLAT.	60 PM / DAMAG.	50 EFFIC IMPROV.	40 COMFOR

REASON FOR WRITE-UP	FM	CM	RM	URM	CL	GN	OTHER	DOWNTIME REQUIRED: Y N

SYSTEM:	CHARGE-BACK ACCOUNT?:	REQUESTED BY:

DESCRIPTION OF WORK REQUESTED:

SKILL LEVEL	UNSKILLED	MAINTENANCE PERSON	LICENSED TRADES	ENGINEER OR OTHER	CONTRACTOR:

LABOR ESTIMATE:	MATERIAL REQUIREMETNS:

ESTIMATED BY:

DATE	INIT	TIME	DOWN TIME	MATERIAL	DATE	INIT	TIME	DOWN TIME	MATERIAL

TOTAL DOWN TIME	TOTAL DOWN TIME

TOTAL (HRS1		+HRS2		)*CHGRT		+(MAT'L		MAT'L		=

ATTACH PAYMENT HISTORY FOR CONTRACTOR

Figure 19

takes found (from the work order header), and send the information back to the people who have the authority to update the asset records. For example, the request that the mechanic verify the nameplate information (such as manufacturer, model, and serial number) against the header information on the work order. It is essential to correct the record when a mechanic discovers a discrepancy or at least start and conclude an investigation (in a very short time).

Asset number: All assets in the plant have unique identifiers called asset numbers. The number is unique, so it is the best way to describe the machine that needs attention. There are different names for this field including Functional Location, Maintenance Worthy Item, tag number, etc. They are all the same in that they identify the asset uniquely. Of course, every asset needs to be clearly marked with its asset number to avoid mistakes!

Location: The location of the work should be included on all work requests. In some instances this is the location of the asset itself. In others it is the location on the asset (for a large asset). This information eliminates the problem of the mechanic going to the wrong area to do a repair. Location is where the unit is that requires work. An address or building number should be included if required. This information might be in the asset master file (when you enter the asset number for the big press, the system looks up the big press and finds the location –Press Room near location B5)

Date/time, Date Opened, Date Closed: The date should be when the dispatcher wrote up the form or entered the request. For user write-ups, the date and time should be when the request was received by the maintenance department. Many systems time/date stamp all incoming work requests. This is an excellent idea since one useful benchmark is the average response time for each priority work. The common benchmarks MTTR (Mean Time To Respond) and MTTR (Mean Time To Repair) are added together to produce the Mean Time from request to completion (sorted and looked at by priority).

On standing orders the date opened is when the work order was initiated and the date closed is when the work order is full. Usually a new one is opened at that time. A standing order might stay open a whole week, a month, or longer, depending on how the information is used.

Time down (1*), Time released to production (4*): Numbers in parentheses refer to chart below.

If downtime is crucial then there are 5 times that are important. The first time is the accurate time when the asset went down (1). The other times are when the request came into maintenance (2), when the mechanic got to the machine (3) and when the asset was ready for production (4) and finally when good production started again (5). With all five times in hand, analysis of response can proceed.

Common benchmarks are MTTFR for each priority of work (Mean Time To First Response 2-3) and MTTR (Mean Time To Repair 3-4).

The time that the request was delivered to maintenance (2) is important to prevent games being played, such as blaming maintenance for excessive downtime. Production systems assign a cause for all downtime. In some instances, maintenance might be charged with 8 hours (1-5) of downtime when they were only brought into the loop after the asset had already been down for several hours (2-5).

Time available for production (4) should be when the unit is completed and able to produce good quality parts at regular production rates. The second problem area is when there is a significant wait after the asset is repaired (4) and when operations start it up (5) (if, for example, they don't have an operator available).

The definition of these times is an item of discussion between production control, production, and maintenance. These differences might be a significant source of conflict.

User, Phone, E-mail: Unnecessary hours are spent each year trying to get access to equipment, units, and locked rooms. In another example, the mechanic may need to talk to the operator or production supervisor to clarify the requested service.

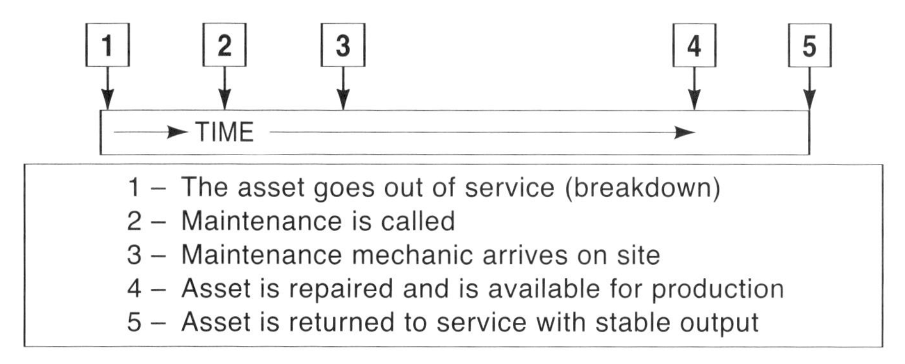

Figure 20 Critical Times; Timing of Work Orders and Downtime

Priority: Priority helps assign work where there is more work than people, and it insures that vital work is not overlooked in the rush of urgent (but unimportant) jobs. Priority systems have a habit of being abused so that users can get their (lower real priority) work done faster. Typical priority codes include:

100. Fire, safety, health (clear and present danger with automatic overtime or contractor authorized until the hazard is eliminated).

90. Statute, operating license or Code violation, OSHA violation, EPA discharge, overtime authorized

80. Breakdowns that stop production, overtime authorized

70. Fire/Safety/health (potential danger to user, public, employees or environment) Overtime NOT automatically authorized

60. PM activity, Potential breakdown including core damage, or loss (all types of minor leaks, deterioration that will get worse)

50. Efficiency improvement, machinery improvements, project work, re-engineering

40. Comfort, Change of Use

30. Cosmetics

Mechanical Priority Systems

For a long time, large organizations sought a system where the computer assigned priority to incoming work. This arrangement would avoid the maintenance department being the scapegoat for priority decisions that didn't go well. It also would avoid the operations problem. Left up to operations, the priority would always be HOT, RUSH, 1, AAA or ASAP. A mechanical system would look at the job, the asset, and apply a predetermined formula to determine when the job should be done. In the best examples, the formula was the result of meetings between operations, process engineering, maintenance, and other groups such as safety, quality, compliance, etc.

One well known mechanical system for prioritization was known as RIME (Ranking Index for Maintenance Expenditures). This system involved multiplying a number of factors together to determine the priority of the job. Another, similar system is recommended by Mike Brown of NSI.

In the NSI system, the criticality of the equipment and the date promised affects the priority. The factors are multiplied so the higher the priority number, the lower the priority. Three, four, or five factors are considered in mechanical determination of priority. The factors below are adaptations of the NSI scheme.

To determine the priority, multiply the factors.

You will notice that if you are already in violation of the law, or there is a clear and present danger, the priority is 0 or the highest. If the job is overdue, then the priority is increased (in some versions the priority goes up further if the job continues to be overdue).

By using different numbers you can fix the priority system to force work that you determine is important but not urgent (such as PM or Corrective jobs). The manager or planner can manually change the priority to suit that day's manufacturing needs.

Factor 1 (Machine or process criticality)
You would pick the values and assign them to each machine or area.
For example:

1. Failure would shut down the whole factory
2. Failure would cause a significant drop in production from the whole plant
3. Failure would shut down a whole section of the plant
4. Failure would cause a significant drop in production from the whole section of the plant
5. Failure would shut down a line
6. Failure would cause a significant drop in production from a line
7. Failure would shut down a machine
8. Failure would result in reduced production from a machine
9. Failure will not have an impact on production

Factor 2 (severity)

1. The asset is currently broken
2. The asset will be broken within 24 hours
3. The asset will be broken within 7 days
4. The asset will be broken within 30 days
5. The asset is not in danger of breaking soon

Factor 3 (Type of Work)

1. PM and PCR (Planned Component Replacement
2. Breakdown
3. Corrective maintenance
4. Efficiency improvement
5. Routine work
6. Cosmetic

Factor 4 (Safety or environmental consequences)

0. Clear and present danger or currently in violation of license or laws
1. Potential Danger or potential violation of license or laws
2. No probable danger

Factor 5 (date due)

.5 The job is overdue
1. The job is due
2. The job is not yet due

Job List	Criticality	Severity	Type	Safety	Date due	Final Priority
Main transformer arcing	1	2	2	0	1	0
Packing machine dead	5	1	2	2	1	20
Main Conveyor chain PM	2	5	1	2	.5	10
Repair and clean sink in bathroom*	9	1	6	2	.5	54

**- RIME methods don't always result in low priority, but important jobs are done in a timely way. You can either make the method more complex or manually modify the priority so that this type of job is done quickly. Usually a bathroom problem should be addressed pretty quickly (depending on the circumstances)*

Reason for Write-up (or Repair Reason)

Code	Description
	Plannable work orders (can be planned and scheduled)
PM:	PM (preventive maintenance) task list activity such as inspection, lube, adjustment and Survey (an initial PM inspection).
CM:	Corrective maintenance (includes plannable maintenance known in advance, when the PM worker finds a potential or impending problem)
SW-	Safety walk-through
UM-R:	User maintenance - Routine work or standing work order (familiar work done every week)
UM-P:	User Maintenance - Project work requested by production (usually small jobs, can be planned) Larger projects are considered RM type maintenance.
RM:	Rehabilitation maintenance, rebuilds, capital projects from management deci sions RM-M: Modernize equipment to shop specifications RM-I: Installation of new equipment RM-E: Efficiency improvement RM-U: User-initiated modification
PSS-	Personal service scheduled in advance such as errands, pick-ups, etc.
CL:	Cleaning machines and shop, TPM cleaning, sweeping up, etc.
	Un-plannable Activity (Must be dealt with immediately)
BR:	breakdown (requiring immediate action). BR could be jam-up, smash-up, crash-up, slow down, leak, quality problem, immediate safety danger, etc.
PS:	Personal Service, errands, pick-ups, minor jobs around office needed immediately
D-R:	Reported damage (someone made a mistake, broke something and reported it)
D-U:	Unreported damage, no report, includes vandalism, sabotage.
OB:	Other breakdown, including code violation, safety audit, OSHA inspection, PM inspector finds imminent danger or breakdown (cannot be scheduled)

What is the source of your work? Sometimes this is a critical question. If you have planning and scheduling in place, and what you think is an effective PM program, and the bulk of your jobs come from breakdowns, then something is not right!

All work requested is requested for a reason. These reasons should be noted on each work order for future analysis. Reasons for repair codes should be built into the code structure of the CMMS. Remember, when you pay a mechanic or a contractor you are buying your maintenance history. Repair reason allows you to analyze your demand for work.

If you were a profit-making maintenance shop (such as a maintenance contractor or a repair shop) you would want to know where future work was coming from. If you found that 85% of your work load showed up on your doorstep every morning you could optimize your business for that factor. The same holds true in maintenance departments except that you might try to change the factor itself.

On the previous page you will find lists to help you see more clearly that in most systems you can specify your own codes for unique repair reasons. and possible reasons to initiate a work order also some data from the chart of hours of the reasons for repairs. (Refer to observations chart on the next page.)

Observations

- ✓ Unplanned events of all types were more than 50% of workload in the first year, almost a third in the second year and down to about a fifth in the third year.
- ✓ Breakdown hours didn't significantly change until the third year.
- ✓ Corrective maintenance went from zero to almost 25% of the work load in the second year. If you add in projects and management decisions that were plannable the workload went to 43% in the second year and 58% in the third year
- ✓ About 4670 net hours were needed to pay for PM system in second year
- ✓ PM and Safety walk through started in second year
- ✓ Safety program went into place with the PM effort perhaps mid-year. We can't determine from this chart if it was effective.
- ✓ OSHA immediate drops over 50% in year two and an additional 75% in year three
- ✓ Damage dropped about 20% in the second year and an additional 50% in the third year.
- ✓ Routine work dropped about a third in the second year and an additional 50% in the third year.
- ✓ Operations complaints dropped a small amount the second year and dramatically the third year

✓ Small project work and management decision work almost doubled in the second year and rose another 25% in the third year

EXAMPLE: In a recent review of Repair Reason in a manufacturer's maintenance shop we found the following hours:

Reason for repair hours	2002	2003	2004
Plannable and scheduled hours			
1. PM/PdM activity	0	600	980
2. SW- Safety Walk through	0	80	160
3. CM-Corrective maintenance	0	4978	2695
4. UM-R Routine work	4706	3245	1675
5. UM-P small user projects	1200	1925	1975
6. RM- Management decision	1023	2080	3021
Total plannable	**6929**	**12,908**	**10,506**
Un-plannable and not scheduled hours			
1. Vandalism/damage			
A. D-R Reported damage	120	240	290
B. D-U Unreported dam	1810	1585	527
C. Other damage	690	345	267
2. BR Operations complaint	5970	5350	1556
3. OB- OSHA immediate	611	240	58
Total unplannable	***9069***	***7360***	***2698***
Total	**15,998**	**20,668**	**13,204**
Ratio of Plannable to unplannable	43.3%	62.5%	79.6%

Figure 23

The great thing about repair reasons is that you can draw conclusions from very little data and (if you are computerized) not too much extra work. In our example it appears that the company added a PM system in year two and added some emphasis on safety mid-way through that year. A lot of work was invested in the PM/Safety effort. In year two, small projects, management decisions and corrective maintenance were mostly probably bringing assets up to standards for PM. These extra expenditures are the funds needed for a successful PM effort. PM systems are usually the result of management wanting fewer breakdowns. In this instance the investment in corrective, management decisions, and small projects preceded the reduction in breakdowns by a year.

In other words, expect a year of lag before good PM practices show up in breakdown metrics.

In the example discussed, it seems that routine work was a catch-all for hours not collected, for uncontrolled work, and for general hiding out. Notice that, as the PM system took hold and the culture shifted, the routine work dropped. That same culture shift accounts for the drop in damage, particularly unreported damage.

The management decisions and small projects stay high because of the shift in culture. Rather than wait for problems to come to them, the company is deciding to go out and tackle problems.

Description of Work Requested

It is important to train your requesters to report the observed condition with as much supporting detail as possible. People tend to blame machines, systems, and component if they have failed in the past. They also tend to blame the part of the system that they are least comfortable with (that is, don't understand). These factors can mislead the mechanic. Instruct the requestors to note on their request what was happening just before the breakdown.

Example:

A process control company whose systems controlled oil terminals received a call (in the middle of the night, of course) that the main computer had shut off and the blending sub-system was off line. The mechanic arrived at the terminal and spent several hours doing a complete system analysis. Everything seemed fine except that the output was failing to start the pump on one of the products being blended. The mechanic also noticed that the gauging system seemed out of range. The second product's pump was not being activated because the source tank for the first product was empty! The system was acting exactly as designed.

The "work requested" described a broken main computer. The technician responded to the description and was misled for several hours. Of course, a more experienced technician (or one with more sleep) would have checked the assumptions before jumping in. Good descriptions of work report to the maintenance department only the observable rather than the inferred information.

This axiom is critical:

Remember, the better the request, the less likely a wild goose chase will follow.

It is a good thing if many people in your facility feel comfortable enough to generate work requests (or notifications on SAP). Having a wide group gives them more ownership and helps them feel a part of the operation (as opposed to reporting everything to the supervisor, who decides whether or not to pass it along).

It is a bad thing if all these people have access but do not have the faintest idea what fields are important, essential, preferred, or nice to have. It is your responsibility to design a mini-course in how to fill out a work order on your system. There are two aspects of what buttons to push and what to put into each field and why. Include both aspects. This information transfer could be in the form of a Power Point, Intranet presentation, or live training with an instructor.

If you design your own work order be sure there are separate fields for work requested and work performed. This arrangement will ensure the integrity of the database by keeping the customer's perception level out of the history file as a cause or effect. We want to collect data on the work requested for another reason. At some point we will be able to build a table between the job requested and the job performed. We can then give the mechanic the highest probability problem based on what was requested (independent of whether or not the requested work is 'right').

Some CMMS include a code for System, which is an artificial category that helps facilitate analysis. This code describes the part of the asset that will be worked on, if known. Systems might include electrical, hydraulic, power transmission, controls, motors, etc. Specific examples of systems would be a circulating pump on a boiler, a hydraulic pump on an injection molder, or a lubrication injection system on a mechanical punch press. The definitions are unique for each class of asset and artificial categories. In one plant, the boiler circulating pump might be part of the system that includes chemical injectors, low/high water cutoffs, and even other items attached to those pipe networks.

The system is broken into a series of codes. The system code helps the maintenance analyst look for root causes. For example, using the system code and the elapsed machine time between failures (MTBF) we might be able to detect a negative trend (MTBF is declining). Without system codes it is difficult to know if the failures are related to each other (since you would have to read each record).

Idea for action:

Most of your service calls concern assets that have been a problem in the past. Given that 90% of more of all maintenance work is repeated, you can prepare for calls ahead of time. If someone calls in with a problem, have a list of questions to try to filter out the non-maintenance and frivolous requests. Be sure that a copy of the questions is reviewed during the customer/user orientation. Using this procedure could cut the number of service calls. Include `not plugged in' problems, procedures for overloads, material outage, wrong tooling/materials.

This approach is both a method of triage and of encouraging self-repair of simple and common problems. It was pioneered by Xerox Corporation when they first started putting large copy machines in business settings. They called it the 15 Questions. The goal was three fold. They wanted to know as much as possible before they dispatched the service person, they wanted to eliminate frivolous calls, and they wanted the customer to be very happy with the machine and with Xerox.

Early copy machines were the most complex equipment ever put into a business setting. In the beginning (and still true with large systems), Xerox set up the machine and trained a key operator. The key operator literally had the key to the machine. As part of the training, the service technician showed the key operator a sheet of questions that would be asked when a service call was made.

From reviewing service records and (needless to say) well-written work orders, the service analysts knew all the reasons why people called for service. Fully 15-20% of the service calls were for things that the customers could reasonably be expected to deal with themselves such as no paper, unplugged unit, or documents left in the feeder (or on the glass).

The 15 questions were designed to provide information to the service technician. They also were worded to lead the key operators to fixing simple things for themselves. The questions were designed carefully to avoid making the key operator look like an idiot. In other words, they did not ask if there was paper in the machine, instead they asked what kind of paper was in the cassette. They didn't ask if the unit was turned on, instead they asked which indicator lights on the panel were illuminated.

Based on the answers, the technician had a pretty good idea of what was wrong. The big bonus was that 15% of the problems were fixed by the key operator. The secondary bonus was that the key operators who participated in the program were more positive about Xerox and their copiers. The key operators reported that they felt more in control and more effective. As a result, they went through the list themselves before they called in and then didn't have to bother the service call center.

Variations of this 15-question strategy are appropriate in many maintenance settings. Many maintenance departments now have a series of questions like these to ensure that simple problems are fixed by the operators, and serious, expensive, or dangerous problems are responded to by the maintenance department.

Requested By:

The work requestor needs to be authorized and preferably trained in the maintenance request system. Depending on your organization, the requestor will be your system (for PM), you or engineering (management decision), your user (breakdowns and small projects), or staff (corrective maintenance work orders).

Date wanted:

When does the customer want to job to be completed? Companies should have restrictions against unreasonable date requests. One technique automatically authorizes overtime and, if necessary, authorizes a contractor. The best arrangement is for the overtime to be authorized against the requestor's budget.

Charge-Back to account:

The accounting system might require that different types of repairs be coded or charged to different accounts. Also, many service requests are paid out of the user's budget. These charge-backs need to be well documented because some of the charge-backs will be questioned. In general, where charge backs are used the PM work is not subject to discussion.

Skill Level

Certain jobs require higher skills or special licenses. The manager/planner should evaluate the skills needed. Some localities require licensed trades for certain jobs. If a contractor is used, the work order should be attached to the Purchase Order. The same information is required for in-house or contracted jobs. On larger jobs this process becomes more complicated because many skills will be needed in a particular order. In that event, the skill level on the work order would be superseded by a labor plan for the job.

Labor estimate, Material/ Tool/other resource Requirements

This block should include a description of the work to be done, broken into logical steps. The cost of each step is estimated. The labor estimate is determined by experience, observation or study. The tool requirements alert the technician as to any special tools are likely to be needed. This block should include a description of the estimated amounts of materials to be used and their costs. Small supplies would be included. All major or recurring parts should be included. Other resources such as lifting equipment would also be called out. These requirements might be superseded by provisions in the planned job package.

Chart showing Information in the Header of the Work Order

Data Field	Source
Asset number	Requestor
Location	Either requestor or master file for asset.
Date/time, Date/time Opened,	
Date/time Closed:	System dates
Downtime	Decide ahead of time which assets are critical and should be tracked
Time down	From production system or from mechanic
Requested By, User, Phone	User and requestor might be the same person or not. Be sure the number or E-mail address is valid
Priority	Decide ahead of time. Could be a calculation, or assigned by planner or requestor
Reason for Wrie-up (or Repair Reason)	Assigned by dispatcher or planner
Description of Work Requested	Requestor
Date wanted: Charge-Back to account:	Requestor Requestor or look-up table
Skill Level	Planner
Labor estimate, Material/ Tool/	Planner
other resource Requirements	

Body of Work Order (information known only after repair is complete):

Time in, Time out, Hours

The mechanic should write the time they arrive, leave and the hours on the job. If mechanics leave a job for any extended period they should clock out. They clock in when they return to the job site to complete the work. Of course, the extended time away should be charged to another work order, a log sheet or some other place. Use the same standard for expressing time (like hours/minutes). One issue is to define whether lunch, breaks, and excessive time to get parts, should be charged to the job.

Date, Initials, Time, Down Time

Work specified on the standing order will often be done in stages on different days and logged to the same document. Each time work is done, the date, initials of the mechanic, the elapsed time on the job itself, and the total elapsed time the unit is down (when `Downtime Required' block is checked), are recorded.

Description of Work actually completed

This description of the completed work is of critical importance. The mechanic should write a quick description of the work actually done. This description might be radically different from the work requested (which is ok), but it should be rich enough to reconstruct what happened. Keep it short. If you go through your closed work orders and see a lot of comments like Fixed, or OK, as the only feedback, then you have a problem in this area. The other essential aspect is that this field must be captured and entered into the CMMS. In the best organizations, any data such as sketches, diagrams, or dimensions are also captured.

Task, Materials

On a string PM work order, each task is to be written out, with a list of the materials needed. Attach extra sheets for longer strings. Keep in mind that string PM is usually a few simple tasks on several machines.

Quantity, Parts, and Materials, Prices

Whenever parts or materials are used they are recorded on the work order in these columns. Include the total price for all parts used. When an item is drawn from stock then you have to look up the price and insert it (many CMMS will do this for you). Put in the part number where replacement parts are used if there is a possibility that you might need to use the number again.

Outside Services

This field is for any services such as rebuilding cores, prefabrication, NDT, outside engineering, etc. On some jobs, equipment charges such as crane rentals are also recorded here. Any other charges that will come from the outside also are recorded here.

What was found: notes from mechanic, mechanic's comments

This field is related to the Description of Work actually completed. Both should be used together. Frequently the mechanic fixes something or finds something not anticipated by the work requestor (for example, a broken pump call results in replacement of a power supply on a PLC rack). This field allows the mechanic to feed back information in addition to the field work actually done. It is essential that details about what was really found written here become part of the permanent record for the asset. These notes are essential for root failure analysis.

Date/time Completed, Inspected by

On some jobs the work has to be closed out by an inspection. The inspector can be the mechanics themselves, an inspector, a satisfied user, or the supervisor. In a major oil refinery, the PM crew signed off on all repairs before the unit could be returned to service. The justification was that life safety was such a big issue and the PM crew was trained in safe return of equipment to service. In this refinery the PM crew was also an elite group in terms of years of service, skills, dedication, and pay.

The body of the work order

The body of your work orders forms the history for which you pay each month (through payroll, contractors, parts and materials), and forms a unique and valuable history of your plant. Many people think that the extra 0.2% (or less), needed to complete the work order and enter it, is money wasted. However, it is money very well spent if it provides the information in a form that can readily be analyzed. The answers to the questions that can be asked are potentially worth 100's of times the cost of acquiring the information because they can impact production output by eliminating repetitive failures and speeding diagnosis and reducing downtime.

How to Audit the body of the work order

The audit will reveal what is important in your environment. Add items that you feel are important to the audit and training class. One example is the need for accurate arrival time for the mechanic. In some plants, the time the request was sent to maintenance and the time the tradesperson arrived is an important metric. A process (such as writing the actual time of arrival and having an operator sign it) should be agreed upon for how to collect the information and included in the work order training.

During this review, create a check sheet adapted to your system based on the information you expect the tradesperson to fill out, like the one below. The course should follow elements of the audit except that more time should be spent following up problem areas.

Number of Work Orders Reviewed (be sure to choose them randomly):		________ = Total		
Date Reviewed	Reviewer			
Review this field	**For this information**	**Checkmark if the work order fails**	**# Fails (F)**	**F/Total**
What was found?	Concise information about what caused the work requested.			
What was actually done?	What was done to fix or mitiga te the problem found? Should be shortand follow plant conventions			
Materials actually used	Correctly identify materials and add any free issue items.			
Special Tools and equipment used	Correctly identify materials and scaffolding.			
Outside services used	Note any outside services used on this work order			
Time in-out 1	Elapsed time must be accurate to +15 minutes.			
Time in-out 2	As far as practical correct the job steps and update estimates for each step			
Time in-out 3	If downtime is important the time when the unit was returned to operations should be noted.			
Notes from trades person	Is there any information that should enter the permanent record? Anything that would help the next mechanic?			
Other________				

Depending on what you find, consider training if problems exist. Only relatively few people are generally involved. Coaching or conducting classes from a pre-designed format is in order. The format should be designed with basics and can be supplemented as time goes on to reflect the increase in sophistication.

Curriculum: Training course for completing the Work Order

The key to effective future analysis is in the information from the mechanic. The important question is, how many mechanics realize the importance of their notes and entries on the work order? One effective teaching method is to ask the mechanics to do some basic analysis from the existing data. They might find such analysis difficult because of problems with the data integrity. This course can be based on solutions to those difficulties.

The very simplest questions cannot be answered (questions like how much did we spend maintaining this machine?), even with the mechanic's cooperation.

There are three issues.

1) The completeness of the data entry
2) The accuracy of the data entry
3) The consistency of the data entry nomenclature (the same thing must be called the same thing each time)

Summary of Work Order is a summary of costs and elapsed downtime.

All calculations are done after the job is completed and the work order is turned in.

Totals

If you are using a manual work order, add up all the hours and write in the totals. If you use a contractor, put in their charge out rate (if known, otherwise put in the total dollars for the contract that is related to this work order). Extend the material total and add that to the labor total for a work order total. Add the totals for any rentals, rebuilds, or outside services. Every work order should have total charges for everything, just like an invoice from an outside company.

With accurate totals, business decisions are easier to make and more accurate. One area of interest is that some systems also keep the hours in addition to the dollars. This procedure is useful because the hours won't inflate but the dollars will. In ten years, the dollars for labor will be far less useful than the hours of the labor.

Data field	Mechanic Training should discuss at least:
Complete the header course	The tradespeople should have completed the course to fill out the header of the work order. They should be conversant with all the codes and categories used by the system.
Materials actually used	The job plan might call out certain materials. The goal is to get accurate data concerning materials and fix any problems in the planned job package. The job might use some, none, or all of the materials. Additional materials would be added, with notes explaining whether the need was for special extra work or materials that should be added to the work plan in future. Supplies and free access materials should be recorded also..
What was found	What did the mechanic determine was wrong with the system (if anything)? An accurate statement can help future diagnosis because the service requested may then be linked to what was actually found. If that same service is requested in the future, the next mechanic may be advised to look at the same cause (in addition to other causes). This procedure may also be used for MTBF analysis (component life for example) so that a consistent description or code should be used.
Work actually completed	Based on what was found, what work was accomplished? Some systems have work accomplished (WA) codes, which simplify this entry. Some WA codes might be R&R (remove and replace), Repair, Rebuild, Adjust, etc.
Special Tools and equipment	The goal is to correct the planned job package with what was actually needed for the job. Special conditions that caused the need for particular equipment that would not be repeated in the normal or average job should be noted.
Other Resources used	This heading would be the place where the mechanic could correct the plan as to lifting equipment, lifting gear, and special tools (not included above) etc.
Outside services used	Note any outside services used on this work order. These services would include the use of contractors, inspectors, NDT, engineers, and what roles they took.
Time	The total of hours spent on the job is essential for analysis. There must be agreement about how the time for breaks, interruptions, and lunch are handled. What if the mechanic goes off site to get parts? How should that be included?
Time	One goal of the training is to get input from the tradespeople to correct the job plan (if you operate in a planned environment)
Time	Maintenance downtime calculations (time service request came in minus time returned to operations) are based on accurate entry of when the unit was returned to operations from the body of the work order.
Notes and comments	Anything that the mechanic thought was important maybe very important at a later date. Such items might include diagrams and sketches, specialized information, and problems with the job plan, problems with the permitting, in short anything that the mechanic wants to pass on to the next person.

Figure 26 Curriculum for Work Order Training Class

CHGRT (Charge Rate)

The charge rate is the burdened labor rate for your facility. In some places, each level of mechanic or trade has its own charge rate. In others, there is a shop rate that includes everyone. The charge rate should include labor, fringe benefits, overheads, and a recovery factor. The Charge rate is discussed in detail in the chapter on accounting.

In all systems, there are hours that are not charged to any work order. For any year, the recovery rate is the ratio of payroll hours to work order hours. In other words, if 90% of your payroll hours are charged on work orders, then divide the charge rate by 0.9. The 10% lost hours will then be paid for by the work order hours charged.

If a contractor worked on the job then the charge rate is also the amount that the contractor charges per hour. If you have just one charge for the whole job (lump sum), then skip this part and fill in the totals for the Work Order.

TOTALS (from the Standing Work Order)

This row allows you to calculate the total cost of the Work Order. The formula adds the hours from the two columns together and multiplies by the CHGRT (charge out rate) to give the total labor cost. There is room to total both material columns for a total material cost. The grand sum is the cost of the work order.

Repair History Jacket

In the beginning of organized maintenance departments, all the machines had file folders. Work orders had multiple copies, and one copy was filed chronologically. The back copy, which generally had the mechanics comments and any drawings on the back, was filed in a folder for the equipment. Fat files meant a lot of activity. Fixed information about the asset such as serial number, model, specs, etc. were written on the outside of the jacket.

Today the information in the asset jacket, book, or folder, can be kept on the computer. Some organizations also keep the folder in paper form because of all of the little sketches and as-built doodles. Many organizations have document management systems that could easily be used to store these additional tidbits. The physical location is less important than the ease with which you can get to the data.

If there is a physical file folder it should contain:

- ❑ Survey sheet (one time or periodic review of equipment condition)
- ❑ Photographs of record over time (with dates)
- ❑ Equipment record card or contents of the equipment master file from the CMMS.
- ❑ Print-outs of all work orders (working copies with notes if possible).
- ❑ Copies of all engineering data relating to unit
- ❑ Wiring diagrams, as-built drawings, documented modifications
- ❑ Planned work, shut-down work lists, past shut-down lists
- ❑ PM list
- ❑ Planning guides for various repairs, feedback on results of repairs
- ❑ A wish list of things you would like to change about the asset.
- ❑ PM justification (why the unit should be on PM)
- ❑ Justification for each item on the PM list based on history, experience, and economics.
- ❑ Copy of Recap sheets for the asset.
- ❑ The folder might consist of several physical files. The main file should contain an index of other material available.

The files should be located in the maintenance technical library.

Maintenance Process Aids

Good maintenance processes are time consuming. Without process aids it would be impossible, in today's slimmed-down factories, to process work orders, recap costs and incidents, and produce reports manually. Three or four decades ago, that was the only choice. Today we have help.

Computerization of Maintenance CMMS (Computerized Maintenance Management System)

The reason why we computerize is the same as why we manage maintenance in the first place. We computerize to lower or avoid costs, improve service, control costs, insure up-time, improve quality, etc. We also computerize because running manually looks bad in the eyes of our peers and ourselves (called the 'because factor' by Jay Butler in Maintenance Management).

Some firms computerize for the last reason because maintenance is the final frontier (the last department of the organization that is still manual). It is sobering to see the maintenance managers of some august high-tech organizations explain that they cannot get adequate funds for specialized software to help their effort (of course they are free to create any Excel spread sheet they want). This attitude reinforces the belief that maintenance has a very low priority and cannot get attention or resources for improvement.

Many maintenance departments are grappling with the decision to computerize or re-computerize (replace an outmoded and under-utilized system). The decision is only the surface of a much deeper decision. Choosing to computerize is also a decision to treat maintenance as a serious profession. The decision to computerize is also a determination to impose discipline on a group of people who are very independent and traditionally hard to control. The computer is a tool that maintenance managers imagine will allow them to predict, effect, analyze, and eventually control what goes on in maintenance. Make no mistake about it, this computerization decision and the deeper decisions that it represents go to the core of the culture of maintenance in any facility.

One trend is the installation of enterprise-wide management systems such as SAP. These installations can cost hundreds of millions of dollars and in most factories the Plant Engineering/maintenance module is given little thought and little or no funds. Why is management so penurious, considering that maintenance spends more than many other departments and is a great, uncontrolled, cost area?

An article in Maintenance Technology magazine states that 60% of CMMS installations do not generate the return on investment (ROI) anticipated. The reason that most CMMS installations go astray and never realize their promise, is because the inquiry that the system came out of was not wide enough. In other words, the people thought they were only computerizing and asked only hardware, software, and database type questions. They never asked themselves the deeper questions of what we are about, how do we view maintenance, and what is our role in the organization for the next several business cycles. In other words, they missed the questions concerning culture.

In the old model, maintenance people are looked at as fixers/maintainers, not thinkers. It is logical in the old model to assume that the tradespeople have no reason to interface with the computer. Paradoxically, management holds an unrealistic view of maintenance in the other direction too. It would be funny if it was not so costly, but management thinks of maintenance as a bunch of do it yourselfers. So management's prevailing thought is, here is the CMMS, go do it yourself (oh, by the way, get your regular job done too)!

In fact, maintenance leadership time is clogged up with paperwork, regulatory issues, human resource questions, budgeting, quality inspections, meetings, and the problems of scarce resources. They are already working long hours just to keep the ship afloat. There is no time to even read a report let alone deeply study an issue. Using other resources available to maintenance is a joke because the maintenance engineers, planners, and analysts are mostly gone in the cutbacks or if there are any left they are buried in emergency work, regulatory, safety issues, and endless meetings.

Herein is the dilemma of the computerization effort. How we grapple with this problem will, in large part, regulate how well the system is used. There is only one solution to this problem in the new paradigm of maintenance management, and that solution is distasteful for many organizations. The solution is:

- The system is installed with adequate training of everyone in the department. The training is in three waves. The first wave is codes and databases where the choices of codes are made and the database is built (during set-up). The second wave is usage. Everyone is trained in how to use the system, enter work orders, look up data, and generate reports. The last wave is benefits. Everyone is trained in how to benefit from the system and get the ROI promised.

❒ After the system is installed and bedded down, mechanics and tradespeople enter their own data and interact with the system on a daily basis. We should help them in all ways possible, with bar code scanning, rational data structures, and streamlined data entry.

❒ All fixed data entered in the system in the first wave is audited by the mechanics on an on-going basis. This data includes things like name plates, PMs, and job plans. Whenever something new is found, it is reviewed by a system czar and changed, all within a week or less. This strategy is critical in the beginning of system use.

❒ Mechanics and tradespeople know how to and are encouraged to analyze data to uncover and solve problems. We should help them by standardizing types of analysis and teaching pre-designed models.

❒ Thinking and analysis of trends is initiated by a work order that says (something like) Do a Component Life Analysis of Unit 30. The tradesperson who gets this work order knows exactly what to do (from training in analysis protocols) and who to brief with the results.

This view appears to threaten the status quo because it apparently moves the decision making down to the bottom of the organization. It threatens the way we look at our mechanics. In truth our mechanics have far more mental capabilities then we give them credit for.

We must never forget that computerization of maintenance is a complex job when measured against other computerization efforts. The reason it is so complex is in the nature of the data collected. Maintenance data has copious detail and flows in a variety of channels.

Why do 60% of the CMMS installations fail to provide the calculated returns?

The complexity of the culture and difficulty of the problem is never taken into account in the implementation process.

Complex Data that is not always easy to collect:

CMMS require interaction with a myriad of data sources and types. Data collected includes: who called in, time called in, time completed, elapsed downtime, what they reported, why is the repair needed, where did the event come from, who authorized it, what priority, where to deliver the parts, when was the last time this happened, what to service, parts used, supplies, trades, crafts, technicians, hours, crews, bench time, rebuild effort, assets, components, sub-components worked on, what was done, down time, what the mechanic saw or recommends, who will pay, comments from the mechanic, comments from the operator, what cost center, and on . . .

Data comes in through many channels:

Telephone, E-mail, face to face, fax, computer networks, engineering drawings, specifications, gages, MAP systems, PLC networks, cycle counters, store room communication, time clocks, authorizations, locations, machine tags, old history cards, written communication from mechanic/from operator, verbal communication from mechanic/ from operator, evidence from the broken parts, outside laboratories.

These information systems include accounting, engineering, and incident management:

The second reason that computerization of maintenance is so tough is because maintenance information systems are among the most complicated packages that are commonly found in industry. To compound the problem, knowledge about good maintenance practices is not well distributed throughout the organization. In sorry fact, even some of the designers of CMMS are not aware of all the good maintenance practices.

Contradiction

Choosing maintenance software is something of a Catch 22. People from accounting, data processing, or even production, do not have enough expertise in maintenance to be of much help in choosing software. On the other hand, the maintenance department people are usually not knowledgeable about the other business systems with which maintenance must interface.

Lower priority and does not get top billing from implementers

Maintenance is low priority when it comes to attention from data processing. Work on maintenance systems is not usually viewed as mission critical (unless you are a maintenance company). This defect leads to a common situation where the maintenance system implementation is delayed for years and eventually the maintenance department is forced to bootleg a small stand alone system that does not talk to the organization's other systems. Data being generated by the maintenance system is then not available to other departments, and the decisions they make reflect this ignorance. In the present generation, top management demands that everyone move to an enterprise system, again without adequate resources or support.

Cultural awareness is lacking

Most organizations focus on picking the right system vendor and the best package. Successful implementations of computerization also focus on the readiness of the organization to accept the new computerized culture. This new culture needs to be sold to all parties involved and interested in maintenance. Without adequate preparation, the system, at best, will only enjoy a superficial acceptance. To improve acceptance, the training effort must start well before the system is selected.

Selecting and installing maintenance software is a major effort that will require the time and energy of key maintenance players. The best implementations have the support of the mechanics, supervisors, and other maintenance staffers before the system is turned on. To facilitate this development, as many levels of the maintenance department must be involved as is practical in the search for a system.

After choosing a system, with the guidance of the vendor and both the maintenance and data processing management teams, encourage the supervisors and lead mechanics to do the research and help input data into the system's master files. Kent Edwards, Vice President for Four Rivers Software Systems (the organization that markets the iTMS package), has a rule of thumb that no more than 90% of the system set-up should be done by either the vendor or the customer. That insures the expertise of the two groups is commingled. For a more complete discussion, see the section on setting up your PM system, which includes a detailed discussion of the steps necessary to set up the CMMS.

Before you decide on any system, answer these ten questions about your proposed CMMS control structure and implementation plans:

1. Consider whether there is enough time, money, and interest to involve all levels within the maintenance department and other stakeholders in the decision process to buy a CMMS? Is there support from top management to see you through the inevitable ups and downs of the entire installation process? Management support is essential.

2. Sufficient resources for a complete installation are also essential. The resources include training dollars, time replaced on the shop floor, and

computer access One common issue is the computer and network chosen for the CMMS. The computer hardware should be chosen so that the system will run quickly and given enough storage that repair files can be kept on-line from date-in-service until retirement. On the people side, can you get typing and basic computer skills training for your mechanics? Will management tolerate the initial research and keying of files by your mechanics and staffers? Can you get the budget authorization to replace the mechanic's slot on the shop floor by overtime or by a contract worker?

3. After the maintenance system is in operation, will mechanics and supervisors have the training, knowledge, positive attitude, and access to the CMMS to investigate a problem? Is there continuing training in advanced concepts beyond `which key strokes to get which reports' type classes? Is regular time set aside for thinking and using the system for research into problem areas? Do mechanics and supervisors have easy access to the system? Are these devices hardened against the shop environment?

4. Is there organizational willpower to insure that garbage and faked data will be kept out of the system? Another way to put that is to ask whether falsifying a work order to fill 8 hours is viewed as a joke or a crime? Will the data coming out of the system be commonly held by management and workers to be accurate and useful? Are maintenance records to be treated as seriously as payroll or other accounting records?

5. Coupled with getting rid of the garbage going into the system is the commitment to make the system increasingly accurate over time. Master files for equipment are to be checked by mechanics when service is done. Necessary updates are to be done quickly. PM routines are to be checked against actual equipment to verify accuracy. Finally, job packages are to be reviewed periodically to see that they are consistent with actual practices.

6. Does anyone (including mechanics) have the time to investigate repair history to detect repeat repairs, trends, and new problems? Related to #3 above, do they have the training to use the system to answer any questions they may have?

7. Can you and your staff spend enough time designing the system's categories to make meaningful comparisons between like machines, buildings, and cost centers? This process requires two steps. The first step is to have the vendor's trainer conduct a class in the category model of that system and how things are commonly handled. The second step is to fight out the categories that you want to use. It is critical to understand and wrestle the decisions that you make at the early steps in setting up a system.

7. If you have 100 pumps, probably 20 of them create the most maintenance load. This rule of management has tremendous applications in maintenance. It is called the Pareto principle (based on the 19th century work of Vilfredo Pareto). Has the Pareto principle (the 80/20 rule) been taught and used to isolate the `bad actors' (that is to identify the problem machines, craftspeople, or parts). Be sure you understand how to generate these Pareto analyses or exception reports in the system you choose.

9. Will you have the support of a responsive data processing department (or a very responsive vendor)? You will want changes, fixes, enhancements. In fact, your ability to handle technology and sophisticated systems will improve after the first 6 months. Many organizations outgrow their first systems in a year or two.

10. Does the longer range plan include CMMS integration with stores, MRP, purchasing, payroll, CAD/Engineering? The trend is toward company-wide networks. Organizations want everyone discussing a problem to be working from the same data. Of course, migration to a company wide enterprise resource management system such as SAP or PeopleSoft is a major step. Linkages will be needed between the maintenance information system and the corporate information systems, with all the links and hooks that that implies. Increasingly, information systems are viewed as strategic advantages. Access to information makes a major difference.

Are you in the market for a CMMS (or are you replacing an existing system)?

How to look for a System

Shopping for a system is a daunting undertaking. There are 200 or more vendors of software for maintenance. There are an additional 250 vendors in specialized maintenance areas such as fleet maintenance, building maintenance, etc.

Sales people (and this is true for all types of systems) know that no sale takes place unless someone in your organization gets excited about their offering (and that person has to be high enough up in your companies' food chain to make a difference). To create excitement the sales person shows you how to solve real problems with the inquiries and reports provided by their system. You complain about a PM problem, they show you the PM screen that solves that problem. You would like to track the costs going into a machine you are building, they show you cost accumulation reports. There is a problem for you with this approach.

The reason why systems fail in their implementation rarely has to do with the lack of reports or inability to get information out. Such failures occur rather from a basic misfit between the system and the existing business process, maintenance culture, or organizational requirements. For example, let's say your culture requires the mechanics to keep log books and the CMMS you choose doesn't support that format. Without a conscientious attempt to win over the mechanics you might face a rebellion over the duplicated effort. Another example is, if you use a very long asset number for accounting purposes, then the 10 digit field of one of the older systems will not do.

These pitfalls can be avoided by looking at any prospective system using a technique described in the next section. We recommend that if you look at several systems you look at the parts of all the prospective systems in the same sequence.

Maintenance Management Software Design

All business application software, Maintenance Management included, have four logical components. Together these sections are the `system.' The completeness and quality of the system depends on the care, knowledge, and goals of the designers of the system in the four areas. When you are choosing a system, designing a new system, or revamping an older system, consider these components separately.

A maintenance system might consist of several to more then 500 programs. All the programs are linked together to form what you see as a seamless system.

Part 1. Daily Transactions (Look at first): Includes all data entered such as Work Order, packing slips/receipts of parts, payroll information, energy logs, and physical inventory information. A defect in this section of the package is usually fatal. It is usually very difficult for the vendor to repair or re-program this section. The main reason that problems here are fatal is the amount of time your staff will spend facing these screens. The second reason is that defects here will adversely impact all other parts of the system and may limit the usefulness of the system.

Look for: Completeness, quick data entry, logical format, consistent format, alternate data entry paths. Alternate data entry paths include work order, log sheet, standing work order, string PM entry screens. The more flexible the data entry, the better. Look for the ability to incorporate new advances in technology including bar coding, use of smart chips, RFID devices, use of portable data entry devices such as PDA (Personal Digital Assistants like Palm Pilots), and wireless. Increasingly, inputs to CMMS are being generated by machines or other data sources (such as time card information from the time keeping system), direct feeds from the shop floor using MAP (manufacturing automation protocol). Be sure the outputs from your machines are compatible with the inputs to the CMMS. Verbal data entry is starting to show up in selected venues, as well as phone tone data entry (the system asks you questions and you enter the answers into a touch tone phone).

You will spend more time entering data into a maintenance system than anything else that you do with it. Most of the garbage that will plague you will enter at this point. Data elements not collected at this point tend to become problems later. For example, a coal to coke manufacturer did not collect the data element `who performed the work'. As a result, management could not look at rework or call backs to determine who needed additional training.

The second issue is consistency. If three people enter the same repair differently (and the system allows that), the computer will have limited ability to analyze that data. An example would be:

1 Replaced bearing A2
2 Fixed the jangling sound
3 R & R (usually remove and replace) thrust bearing on shaft `A'-,
4 I unscrewed the housing D1, put my hands around the end of the shaft holder and. .

These could all be the same repair. A person rather than the computer would have to decide if they were the same. Any kind of component life analysis is extremely difficult unless the components are coded identically.

The other issue of consistency is system editing. Better systems test data coming in against the master files, or with logic to determine that it is valid. For example, equipment ID numbers are checked to see if they are valid (already in the file), subsystems are checked to see if they exist for that asset, etc. Some systems will not allow you to log a repair that isn't on the master repair list for that asset, such as changing the tires on a pump. Other checks might be that you can't close a work order before you open it, or meter readings go up unless the meter was replaced.

A good system might go through 50 different edits to keep garbage out of the database. Total vigilance is required to do so. Garbage in your database undermines all your work.

One of the rules of data processing is to move the data entry chore as close to the generator of data as possible. Any firm that decides to do so should be aware that their training bill will be high for the first year because 20 or more people will have to be trained. One mining operation with a large CMMS and over 500 maintenance people had to employ data integrity officers. These people read most of the work orders and returned those that were not filled out completely and properly.

Lack of speed is the greatest killer of systems. Clumsy data entry design will cause a revolt among the mechanics or supervisors (or whoever enters the information) and lack of compliance with the system requirements. To test the speed of a system, develop a set of 4 or 5 typical work orders. In most systems the master files will have to be amended to accept your work orders. Observe the difficulty of this process (use the information you gain for the next section). Enter the work into the target systems and count the key strokes and time the response time for the process. This exercise would give you one way to evaluate competitive systems.

There are two things to be aware of in this kind of comparison. The first item is to be careful that you are making a fair comparison between like levels of detail capture. A fast system might fail because it doesn't collect enough details. The second thing is that because the servers are not loaded during a demo, the system could be quite a bit slower in use with a few years of data in the files. Ask to talk to a customer of similar size that has been using the system for a few years, and ask them about speed.

Indications of a speedy design include on-line look-up tables to ease the way when work order data is missing, a copy function for common repairs, ability to recall a prior repair and drop it into the new work order, and generic repairs. Others are speed typing, where you can type the first few letters and have the computer fill in the remainder, field duplication keys, and programmable defaults (with the most common choices as defaults). Other things to look for are speed keys (such as Alt + Function keys) to allow an operator to move around the system to avoid the slow but friendly menu structure.

Part 2. (look at this second) Master files: Master files contain unchanging information about the assets, parts, mechanics, and organization. The master file structure reflects the designer's biases more powerfully than any other part of the system.

Look for: Completeness is the big issue because it is very difficult to add any fields to a master file after it's in use. Not having space in a master file for information that you want to store (perhaps after the system is in use) is a common major difficulty.

If a data element is not in a master file, and it is not collected in data entry, then it is not in the CMMS. If data is not in the CMMS then you cannot analyze it with the CMMS. The most common example of this problem is when you want to look at maintenance activity related to some outside facts such as downtime, shift change, or even contract negotiation (when charge rates should change). If those outside data items are not on the computer network or accessible over the Internet, all analysis will have to be done manually. In large networks or mainframe computers, the data might be in another system. The analysis could then be done with special programming on the computer or in a spread sheet package.

Typical data required for the asset Master file

Typical record for a pumping system (several redundant pumps with controls).

Asset number:	MWI #4
Asset Description:	Water pump system
Location:	Hydro finishing
Manager responsible:	V. Santiago
Manufacturer:	Pacific Pumping Company
Model number:	Series III
S/N:	51B39721
Specs, Elect:	3 pump units tied together through control panel 208V, 3PH, 178 amps max.
Connection to which asset:	Finishing, Process water system, electrical system panel #4
Condition:	One of the three pumps is broken and is out of service, the other two operate in alternate weeks.
Work to be done:	Rebuild or replace bad unit
Estimate:	Central pump quoted $122,000 to rebuild.
Prob. of replacement:	Immediate on 1 unit, 5-10 years on other two.
Vendor:	Universal Plumbing Contractors
Installer:	Universal Plumbing Contractors
Date Installed:	1992
Date in service:	1993
Original Cost:	unknown (part of general contract not detailed)
Warranty:	1 year
Control Panel	Pacific Pump
Model	Series III

Figure 27

The usual process to determine the contents of the major master files has two steps. The first step is to look at your data needs. Make a list of the elements necessary to produce the information you need to manage the operation. The second step is to survey several systems' master file layouts to look for good ideas and additions. The first step could be repeated, after the system survey.

One system, TMS has 105, 64-character fields in the equipment master file. The first 10 fields are reserved by the system for their canned reports and inquiries, leaving 95 large fields to store information. Systems such as TMS are very flexible. The price you pay for flexibility is that the set-up is more complex and requires more effort. You have to design the system! On another system there was a field of 35,000 characters (that's about 8 pages of typed characters) for comments!

Report and screen headings

In better systems there is a master file with all report headings, utilization fields (hours and pieces are never mixed), and screen headings (possible called a heading or table master). The advantage is that you can adjust the language of the system to suit your present language and culture.

Part 3. Processing: The daily transactions are generally processed online, in real time. Processing updates the PM schedule, summarizes detailed repair data for reports, records machine histories, and keeps all financial records current.

Look for: Does it work? Process some data through the full cycle and see if all the accounts, schedules, and master files are updated correctly. Accuracy and completeness are the difficult areas in this item. Most of the bugs will occur during unusual processing conditions.

On a large system, the processing program represented 4 or 5 months of intensive work. After the original programmer left the company, none of the other programmers wanted to venture into the code. Instead of fixing the problems the programmers wanted to rewrite the entire sequence from scratch.

All established system vendors have a well-oiled (and usually well-used) system for accepting bug reports and issuing patches or fixes. Ask the vendor about this process. These communications channels are essential to ensure that all the important bugs are uncovered and repaired.

Make some inquiries from their customers about bug reporting and how long it took to fix them. If you have access to a user group forum (not controlled by the CMMS vendor), ask other users how responsive the vendor has been to bug patches and service packs.

As you would expect, an older system from a reputable vendor should be bug free unless you are using the system in some unusual way. New offerings from any vendor usually contain bugs. Many IT departments don't upgrade immediately when new versions come out. For example when 6.0 comes out they wait for 6.1, which is usually the bug fixed edition.

Part 4. Demands, reports, inquiry screens: The demands (you make) on a maintenance system include reports and screens. Reports should be generated where there is a large amount of data, or where analysis is required. Inquiries should not have to require going to print. Imagine how you expect to use the system, and what questions you want it to answer, then see how the system will behave.

Look for: Many different ways to look at the data, complete basic sets of reports and screens, future ability to alter or add reports/inquiries to suit your changing needs and growing expertise. Ability to export data sets to spreadsheets for desktop analysis. Another useful program (many vendors either include one or recommend one) is an easy-to-use report writer.

The reports and screens are the reason why you bought the system. You've been feeding data and maintaining master files for this payoff. The reports should be useful, not too detailed (with the ability to go to a higher detail level), include the information you need (not results from 10 other divisions), and be easy to read.

An example of useful levels of detail can be taken from the popular financial program Quicken® (Intuit Corp.). You can ask for a summary of the totals of all money spent by category. If you highlight one total and double click the mouse, the detailed transactions that make up that number pop up in a window. Hitting the escape key returns you to the original report. With this method, the details don't overwhelm you but are easily available.

A Canadian maintenance manager complained that his system gave him too much data. Every week he was treated to a 1400-page report of all maintenance information, containing detailed comparisons with other divisions. He referred to three or four pages and occasionally skimmed the report to see what his buddies were doing in his old division. However, resources and paper were thus wasted. He asked to have a summary weekly and for the big report to come monthly or quarterly. Information services reported that they couldn't find the time to make the change.

Fifty Questions To Help Your CMMS Search

Included here are questions to ask yourself and to ask vendors. The answers to these questions will help you to avoid the most common pitfalls of choosing, purchasing, and installing computer control and information systems. You can get additional ideas from both the Maintenance Fitness Questionnaire and the section on installing PM systems.

A full function CMMS should be able to help in many areas. Many organizations purchase systems to solve specific problems. They don't need other functions or don't consider them important at the time of purchase. The following 50 questions will help you focus your attention in the various areas. They are not in priority order.

Work Order section:

1. Be easy to use and allow future conversion to bar codes, RFID, hand held, and other improvements to technology.

2. Classify all work by some kind of repair reason code: PM, corrective, breakdown, management decision, etc.

3. Provide an easy way for a single person to screen work orders entered before authorization that work can begin. Some systems have a field that has to be checked by a supervisor or manager to release the job to the next processing step.

4. Print up-to-date lock-out procedures automatically. Have the ability to access generic lock-out files and incorporate the right lock-out scheme (there might be only 10 variations for the whole plant). Less desirable (because any changes in the law or company practice would necessitate changing all lockouts one-by-one) but still OK, would be an individual lock-out file for each machine.

5. Automatically cost work orders. Be able to look up the value of a part in the inventory and bring the cost across to the maintenance work order. The system should also look up the charge rate for the individual mechanic. The system will incorporate costs on purchase orders and outside services.

6. Provide status of all outstanding work orders. Allow sorts on different status codes. An essential example would be to print or display all work orders in the Ready Backlog.

7. Record service calls (who, what, time stamp, where, how) that can be printed in a log format.

8. Allow production to find out what happened (what status) to their work request (over the network) without being able to make changes.

9. Calculate backlog of work and display it by craft. Calculate the backlog by

type such as ready, total, waiting for planning, etc. Should have the ability to sort and re-sort the backlog of work orders by location of work, craft, and other ways.

10. Can display or print both open and closed work orders very easily. Keep work orders available for at least 5 years, and preferably from birth to retirement of the equipment.

11. Facilitate job planning with labor estimates by task. Keep task dictionaries, generic (already planned) work orders, provide ability to recall a repair, to make it a master for future use, and to reenter a work plan and clean it up.

Stock room section

12. Facilitate all types of analysis including big ticket analysis by printing all parts over $500. Facilitate A-F analysis by printing the product of (in descending order) the unit cost times the annual usage (to identify parts that consume the most money each year).

13. In the store room part of the system, have part location to help the mechanic or storekeeper find infrequently-used parts. Have cross-references in the location system to find all the storage locations for the same parts stored under different numbers.

14. Generate a parts catalog by type of part or by current vendor with yearly usage to facilitate blanket contract negotiation.

15. Recommend stock levels, order points, order quantities based on usage. Keep monthly usage statistics for at least 24 months.

Maintenance History and reporting

16. Maintain maintenance history sufficiently detailed to tell what happened years later.

17. Provide information to track the service request - maintenance work order issue - work complete - customer satisfied cycle. Include elapsed time for each step and other analysis factors.

18. Provide reports for budgets, staffing analysis, program evaluation, performance.

19. Provide information for work program, work planning, scheduling, and job assignment.

20. Be able to isolate all work done (sort, arrange, analyze, select, or list) by work order, mechanic, asset, building, process, product, division, floor, room, and type of equipment or asset.

21. Provide the ability to easily structure ad hoc (on the spur of the moment) reports to answer questions that come up without the services of a programmer.

22. Have the ability to generate equipment/asset history from birth (installation, construction, or connection) to the present with data for all major repairs and summaries of smaller repairs.

23. Provide reports that will identify the few important factors and help you manage the important few versus the trivial many.

24. Report on contractor versus in-house work. Be able to track contractor work in as much detail as in-house work.

25. Provide reports charging back maintenance cost to department or cost center.

26. Provide reports with mean time between failures (MTBF) that show how often the unit has failed, how many days (or machine hours) elapsed between failures, and the duration of each repair (MTTR).

27. Highlight repeat repairs when a technician needs some help.

PM section

28. Allow mechanics to easily write-up deficiencies found on PM inspection tours. Once entered by PM inspection, system automatically generates and tracks a corrective maintenance work order.

29. Produce PM work orders automatically on the right day, with the right meter reading, etc. Be able to sort work orders by location to minimize travel time. Accelerate PMs that are about to happen so as to schedule them before shutdowns.

30. Be able to display PM workload based on estimated usage for a future period such as a year by week or month, by trade.

31. Be able to record short repairs done by PM mechanic, in addition to the PM and actual time spent. Should code short repairs under corrective maintenance if the system allows multiple work items to be accomplished under coding for 1 work order (desirable trait).

32. Be able to support multiple levels of PM on the same asset (such as a 30 day 'A' level and a 180 day 'B' level on the same asset). Reset the clock if the high level is done (if you do a yearly rebuild, reset the monthly PM clock). A resetting feature prevents a 30-day PM coming up a week after a rebuild (unless of course you want it to).

33. Facilitate PM Optimization by being able to print all failures and PM Tasks. Have screens and tools to support PM optimization.

34. Allows the input of data from Predictive Maintenance sub-systems. This ability might include trending, days to alarm, base lining, and comparison with previous readings.

35. Maintain the history of each PM task. For example, if you modify a PM task, the old task is archived and the reason for the change is recorded. Estimated

Task costs are useful to preserve.

36. Produce simple reports that relate the PM hours/materials to the corrective hours/materials and to the emergency hours/materials? These reports will show the effectiveness of the PM program, and the ratios become benchmarks for improvement.

37. Easily handle a string PM such as a lube route, filter change route.

General

38. Be able to handle 3 to 4 times more assets then you imagine having. Even medium-sized and smaller companies go on acquisition hunts. A small successful manufacturer might find itself tripling or more in size overnight.

39. Have a logical location system to locate assets and where work is done.

40. Be able to track the warranty for components, and flag warranty work to recover funds.

41. Be easy to use and learn for novices and quick to use for power users.

42. Be integrated or have ability to be integrated with purchasing, engineering, and payroll/accounting

43. Run on standard computer hardware (not special hardware incompatible with everything else). The system should run TCP/IP protocol, which allows a browser interface over the corporate Intranet.

44. Be purchased from a system vendor who has the financial strength to complete the contract (and stay in business for several years after installation).

45. Come from a vendor having software support people, with whom you can easily communicate, preferably via an 800 number? Is there E-mail support? Is there on-line chat during business hours? Once you get through, do the people know the product and maintenance of factories?

46. Come from a vendor who provides economical customization? Does the vendor have on-going enhancement and service releases? Are the programmers employees of the vendor or contract workers?

47. The CMMS is the product of a vendor who has a local installation organization, an experienced VAR (Valued Added Reseller), or installation consultant?

48. Be the product of a vendor who is experienced in management of installation projects of the size of your facility. Does the vendor have start-up experience with projects this size?

49. Be produced by technical people who are well cross-trained (in software, hardware, and reality wear, like how a real machine works). It's important that the installation people have experience with maintenance.

50. Come from a vendor who has been in business 5 years or more.

Maintenance Technical Library (MTL)

Increasingly the maintenance technical library is becoming virtual (meaning it exists wherever there is a connected computer). Even with an Internet-based technical library it is great to have a place where maintenance technicians can have access to a computer and printer. The local network or the Internet can then be used to look up repair history jackets, equipment manuals, parts lists, assembly drawings and so on.

One good idea for the MTL is a display cabinet for broken parts, bearings, and other small items. A display where you can pick up and handle burned bearings, fatigued brackets, and other worn items can be a powerful teacher for new workers, plant people outside maintenance, and even maintenance old-timers. Clean them up and label them. You'll find people can't keep their hands off them (but they are learning).

In addition, the MTL should be the location of copies of plant drawings, site drawings, vendor catalogs, handbooks, engineering text books, etc. If you have computerized maintenance, stores, or purchasing, CADD, CAM, then a terminal is located in the MTL.

Considerations

- Explicitly check if back-ups are being made automatically of the CMMS and MTL data files. Normally the IT department is pretty good at back-ups but it doesn't hurt to ask.
- Protect paper files from fire, flood, and theft (consider fire proof file cabinets and off-site storage of copies)
- Use some kind of sign-out system when material is likely to be removed
- Make it someone's responsibility to keep the data up to date.
- Manage the revisions so that all copies are updated (refer to ISO 9000).

When the MTL is set up you will have: a ready reference for make versus buy decisions. You will also have repair history, repair parts reference with history, repair methods referral, planning information source, time standard development, data bank for continuous improvement efforts, and a maintenance improvement team headquarters.

The Internet is now everyone's Maintenance Technical Library (partially adapted from the Internet Guide for Maintenance management by the author)

The Internet is being used for maintenance information in several ways. The advantage of this use is that the capabilities are available around the clock, 365 days a year (as long as you can get access to the Internet).

Find it fast

The search engines have indexes of everything they find as they crawl around the Internet. As you locate useful information you can set your bookmarks (or favorites) to remember the location of any site. If you find critical information (subject to copyright rules) you can download it to your own Intranet so that you know where it is (forever- if necessary).

You can find vendors of everything from valves to engineering services by entering critical search keywords into Google.com, Altavista.com and other search engines.

Catalogs are much cheaper and easier to update if they are on line. On-line catalogs also save your shelf space and trees. Companies can make their latest catalogs available as soon as they are complete. Storage on the computers is inexpensive, so a huge volume of information can be made available such as complete technical specifications, photographs, video clips, audio descriptions, or drawings. All of the data is just a click away. The driving force is advertising budgets. An entire Web site for a year is comparable in cost to a single full-page ad in a leading maintenance magazine. Again, the catalogs that you use are bookmarked for future access.

If you are a shopper for used equipment, the Internet is where the action is. From E-Bay to the myriad of specialized auction sites, good deals abound for smart machinery and parts shoppers. It's easy to locate even the most obscure used equipment and parts: in addition to auctions there are classified ad sections where companies and individuals can purchase, sell, and trade equipment. On some sites, the more exotic items are easier to find (fewer hits at the search engine) than some common items.

What better way is there to distribute Technical bulletins? The latest technical problem and fixes for it can be available minutes after the vendor's engineers decide to put it on-line. No longer is weeks to months lead-time required to publish and mail the bulletins. Also, having the latest information is a click away. The software vendors are light years ahead of everyone else in this area and can give higher levels of support at lower cost through this method. Some manufacturers push the bulletin to you (like the Windows update function when it is turned on).

Drawings, field modifications, and manuals: In the same way that you are updated by technical bulletins, you can view manuals and download drawings. To download means to transfer a copy of a file from the vendor's computer, called a server, to your computer). The file can be a manual, a drawing, just about anything). Wouldn't that be great at 3 am when you can't find the wiring diagram? Field modifications also can be fed back to the OEM engineering departments, if that is appropriate.

The MTL can become a hub of Commerce (once you get purchasing to approve some key vendors and low cost nuisance buys). You can shop for most MRO items from storefronts on the Internet. Major US industrial distributors such as Grainger and McMaster Carr, have a large presence on the Internet that is more detailed and complete than even their massive catalogs. Storefronts cover all types of tools, new and used parts, maintenance supplies, uniforms, and other items. Encryption (a fancy word for a way to scramble transmissions) is becoming widespread to allow high security for credit card numbers and bank information (called secure server).

Rapid parts information and streamlined purchasing are reducing the cost of acquisition: Some sites allow you to look-up part numbers from exploded drawings. You can put your mouse cursor on the part and drag the number to an order form. Add your P.O. or your purchase card information, and ship-to address, and you have placed an order for a quarter of the cost of traditional purchasing.

The MTL benefits from the existence of FAQ (listings of frequently asked questions) files for equipment and materials they use. This facility is particularly useful in the middle of the night, when it is used just before the technician wakes every one up with a basic question. In every field, and on almost every piece of equipment, there are FAQ's. The FAQ file can be read by novices or new customers. FAQ's are on line and available 24 hrs a day, when you have a question.

Technical help directly from the horse's mouth (the OEM engineering department) is sometimes available through the companies' web site. You can ask questions of the vendor's technical departments and get answers back to solve your problems by E-mail or sometimes by on-line chat (similar to Instant

Messaging). Technical departments develop menus of canned E-mails that solve these common problems for immediate response. The technician can then spend time on the more uncommon or complicated problems.

Your MTL can be a portal to a massive library like the Library of Congress. In addition, many university libraries and information databases are available on-line. The library of Congress is putting its enormous library on-line.

Could the MTL be a social place? One social activity is participation in User Groups. User groups are usually organized by the CMMS vendor but not always. If you own a CMMS you might want to talk to other real people using the same system. Many user groups are going on-line as Newsgroups. Here you can read other's comments about the software, ask questions of the whole group, get help, and gripe to your heart's content.

A good technique to get started with improved maintenance management: Survey

The survey is a comprehensive look at all the maintenance needs of a factory or facility. Survey participants are usually organized to go from room to room, area to area, and building to building, reviewing the asset lists and looking at each entry. Part of the survey is to step back and try to see more subtle or more global maintenance exposures (walls, stacks, supports, basements, general piping, etc). These annual inspections are relatively detailed and should be planned when adequate time is available. The survey also looks at areas that might not be covered by traditional PM inspection, and it is designed to be effective for the big picture items for which there is no task list. The survey can be a scheduled activity that takes place before closing the work list for a shutdown.

Write-up any observed deficiencies. These deficiencies become the backlog for the rest of the year. Planned, then scheduled, work is the least expensive, least headache work in maintenance. Almost all deficiencies get worse and become nightmares, given enough time. The key to low maintenance costs is to catch and repair as much as possible before it gets expensive.

1. If available attach: Plot plan or survey (with sizes of the building and lot), aerial photographs, and any sets of blue prints or sketches (plan views) showing the plant floor layout. The plans should also show where utilities enter the property and where shut-offs/disconnects are located. Plan your tour from this set of documents. Include earthworks as well as buildings.

2. Print the master files for each major asset and verify the information in the files. Bring blank asset information sheets
3. Include Maintenance write-ups for all deficiencies (these write-ups will be placed in the backlog file in the maintenance information system)
4. Look at both large and small safety problems
5. Look at all potential environmental impacts including bodies of water, soil contamination, smoke stacks, etc.

Survey process includes:

Must haves:

1. Description of each asset
2. Location of asset
3. Information from the nameplate: serial number, model number, manufacturer, specs

Would likes:

4. Condition of asset, what work is to be done, estimates
5. Replacement value of asset (called RAV)
6. Probability of replacement: immediate, 1, 2,3,4,5 years
7. Copy of the owner's manual, parts list, etc. (Or note of location)
8. Special conditions that would affect PM
9. Digital photographs or video tapes

Inspectors should have the following tools for the initial survey:

Flashlight, Binoculars (for visual inspection at a distance), magnifying glass (to examine paints, surfaces, sub-strata). Other useful items include a folding knife, instant camera (for recording problems or locations, and access to a good digital camera), stepladder, extension ladder, measuring tape, selected hand tools, and flashlight

During an initial survey, any deficiencies should be written-up on the appropriate asset information form. Survey forms follow on the next several pages

Insert Figure 17 on page 110 Asset Information sheet

INSERT Figure 18 on Page 111 ASSET.FR

When you survey the entire facility to look for maintenance and liability exposure, be sure to look at:

Access items such as doors, windows, hatches
ADA requirements (disabled access)
Attachment points
Boilers
Cat walks
Chemical storage
Clean rooms
Communication systems, raceways
Compressors and air delivery systems, vacuum systems, dryers
Computer rooms, shop floor computers
Control rooms
Control systems (like PLC's, MAP systems)
Drain systems, environmentally secure disposal and open
Elevators, people movers
Electrical items (major), electrical distribution systems, transformers, sub-stations
Electrical systems (minor) including lighting, outlets, boxes, disconnects
Environmental systems (scrubbers, separators, filters)
Environmental inspection (asbestos encapsulation integrity)
Explosion proof systems, boxes, inert blanketing systems
Fire fighting systems, fire alarms, other safety alarms
Food service equipment in cafeteria, kitchen equipment, laundry equipment
Exterior finishes, accessories, roofing, roof catwalks, equipment attached to roof, openings
Generators, co-generation facilities, power houses
Grounds, pavement, sidewalks, parking areas
HVAC components (heating, ventilation, air conditioning), exhaust systems
Interior finish, flooring
Legal liability inspections such as fire systems, elevators (use contractors?)
Mobile equipment, trucks, trains, cranes, ships, cars, pick-up trucks, turf equipment
Physical structure of building
Plumbing items (major), pumps, piping systems, rest rooms
Production equipment, process equipment
Quality inspections, certifications, ISO 9000 requirements
Rack systems, automated conveyers, storage/retrieval systems
Safety/security systems: Fire alarm, fire extinguisher, smoke detectors, security systems
Stacks and chimneys
Supports (for tanks, reactors, etc)
Swimming pools, settlement ponds, water intakes, other ponds
Tanks (both underground and above ground), related piping systems, pumps, dryers, filters, chemical reactors, welding, process
Trash compactors, trash handling systems, recycling systems
Waste, hazmat handling systems

Mechanical Asset Information Sheet

MECHANICAL ASSET INFORMATION SHEET	ASSET NAME:		ASSET NUMBER:
ASSET DESCRIPTION:			DATE
LOCATION OF ASSET:		DEPT/COST CENTER	
MANUFACTURER:		S/N:	
SPECS/ELECTRICAL CHARACTERISTICS:			
CONNECTED TO ASSET:			
VENDOR:			P.O. #
OWNERS MANUAL:	DATE IN SERVICE:		ESTIMATED LIFE: YR
PARTS LIST:	COST: $	SALVAGE VALUE: $	
INSTALLATION BOOK/DWGS:	LOCKOUT INSTRUCTIONS:		
CONDITION OF ASSET:			
WORK TO BE DONE:			
NOTES			
DATE ENTERED:	CHECKED:	SURVEYER:	

Asset Information Sheet

ASSET INFORMATION SHEET	ASSET NAME:	ASSET NUMBER:
ASSET DESCRIPTION:		DATE
SIZE:		
LOCATION OF ASSET:		
SPECS/CHARACTERISTICS:		
LOCATION OF DRAWINGS:		
CONDITION OF ASSET:		
WORK TO BE DONE:		
NOTES		
DATE ENTERED:	CHECKED:	SURVEYER:

Maintenance Strategies: Approaches to Deterioration

We face two major problems. As discussed in the beginning of the book, these problems are the globalization of production locations, and the loss of skilled workers through retirement, attrition (or downsizing), and cultural shifts (making maintenance less desirable as a career. We need strategies that will address these survival issues. How do we reduce the cost of producing our products (to nullify the advantages of lower-wage venues) while at the same time reducing the need for skilled workers?

The question is, can we have both low production costs and fewer highly skilled workers, or do we have to give up one for the other? The answer is simply that there are strategies to accomplish both objectives, and PM is one of them.

A properly designed and conceived PM program (with PdM as the inspection component) can reduce the need for highly skilled maintenance workers and improve the capacity of the plant. Improved capacity without the need for substantive additional capital investment reduces product unit cost. PM replaces the need for large numbers of highly skilled breakdown mechanics and electricians, with fewer overall numbers of people including a few extremely highly-skilled analysts, a few moderately highly-skilled inspectors, and a good number of PM foot soldiers (who can also be operators).

The important thing is that the successful system is not blind, deaf, and dumb. Your system has the flexibility and intelligence to adapt to the situation on the ground. Most PM systems overkill with failure modes that are of remote probability, and ignore some obviously dangerous, expensive, or common, failure modes. The result is a massive waste of resources and dubious effectiveness. One powerful strategy is RCM and its descendent PMO.

Reliability Centered Maintenance (RCM)

One of the best models for continuous improvement is the application of reliability centered maintenance. RCM is one of the radical and powerful ways to improve maintenance because it addresses the core of the customer need, that is, an increasingly reliable system. The technology is an outgrowth of deep investigations into reliability on behalf of the aircraft manufacturers and the airlines.

RCM is a five-step process that is usually team driven with members from operations, engineering, and maintenance. Where there is significant hazard, safety or environmental specialists would be included. RCM is usually facilitated by an RCM specialist with good knowledge of the process and products.

1. The first step in implementing RCM is to identify all the functions of the asset. At first this need might seem trivial, but on a second examination you will find many important secondary functions. Functions are divided by primary, secondary, and protective (also called hidden). Each function is defined by a specification, or by performance standards.

   For example, the primary function of a conveyor is to move stone from the primary crusher to the secondary crusher. The specification calls for 750 tons per hour capacity. Secondary functions include containment of the crushed stone (you don't want pieces falling through the conveyor and hurting someone).

2. The second step is to look at all the ways the asset can lose functionality, which are called functional failures. One function can have several functional failure modes. A complete functional failure would be that the unit cannot move any stone to the secondary crusher. A second failure would be that it can move some amount less than the specification of 750 tons per hour. A third functional failure is when the conveyor starts moving more then 750 tons per hour and begins to overfill the secondary crusher.

   Each secondary function also has functionality losses. In our example, the conveyor could allow stones to fall to the ground, creating a safety hazard.

3. Review each loss of function and determine all the failure modes that could cause the loss. Failure modes are not restricted to breakdowns. Operational problems, problems with materials, and utilities are also considered. In our example, the list might be 20 or more failure modes to describe the first functional failure alone. Failure modes on our rock conveyor include motor failure, belt failure, pulley failure, inadvertently turning the unit off, power failure, etc. Each functional failure is looked at, and the failure modes are defined.

Use some judgment to include all failure modes that are regarded as probable by the team. All failures that have happened in the past, in this or similar installations, would be included, as well as other probable occurrences. Take particular care to include failure modes where there might be loss of life or limb, or environmental damage.

It is essential that the team identify the root cause of the failure and not the resultant cause. A motor might fail due to a progressive loosening of the base bolts, which strains the bearing, causing failure. This event would be listed as motor failure due to loosening of base bolts.

It is important to include failure modes beyond normal wear and tear. Operator abuse, sabotage, inadequate lubrication, improper maintenance procedures (re-assembly after service), all would be considered.

4. What are the consequences of each failure mode? Consequences fall into four categories. These categories include safety, and environmental damage, operational and non-operational. A single failure mode might have consequences in several areas at the same time. John Moubray in his significant book Reliability-centered Maintenance says "Failure prevention has more to do with avoiding the consequences of failure than it has to do with preventing the failures themselves."

The consequences of each failure determine the intensity with which we pursue the next step. If the consequences include loss of life, it is imperative that the possibility of failure be eliminated or reduced to improbability.

A belt failure would have multiple consequences that would include safety and operational hazards. A failed belt could dump stone through the conveyor superstructure, hurting everyone underneath. The failed belt would also shut down the secondary crusher unless there is a back-up feed route.

The failure of the drive motor on the conveyor will cause operational consequences. Operational consequences have costs to repair the failure itself, as well as the cost of downtime and eventual shut down of the downstream crushers.

Other failures might only have non-operational consequences. Non-operational consequences include only the cost to repair the breakdown.

5. The final step is to find a task that is technically possible, and it makes sense to detect the condition before failure or otherwise avoid the consequences. Where no possible task can be found, and there are safety or environmental consequences, a redesign is demanded.

For example if it is found that the belts start to fail after they are worn to 50% of their thickness, an inspection might be indicated. If the belts fail rapidly after cuts or other damage, a sensor might catch these problems. Where safety or environmental damage is the main concern, the task must be to lower the probability of failure to a very low level.

Hidden failures

We also have to be conscious of the fact that some failures of sensors or protective devices are hidden. A failure is said to be hidden if it occurs and the operators, under normal conditions, would not notice the problem. For example, if a belt thickness gauge fails (unless the design is failsafe), the operators would have no way of knowing that the sensor is not working. Without protection, a second failure (of the belt) could more easily occur and cause an accident.

In operational failure modes (such as the motor failing) the cost of any PM tasks over the long haul is always lower than the cost of the repair and the downtime. If the PM tasks cost $1000 a year, a breakdown costs $2500, and downtime costs $4000 to repair, the breakdown must be avoided more than every 6 years to break even.

PMO (PM Optimization)

This section is partially adapted from Complete Guide to Preventive and Predictive Maintenance by the author published by Industrial Press

Steve Turner, a professional engineer from Australia and RCM expert, developed PMO out of frustration with the application of RCM in mature industries. He can be contacted through www.pmoptimisation.com.au . RCM is an expensive process, particularly for existing plants. PMO is modified to make it simpler but retaining most of the benefit. PMO is specifically designed for mature industries, where the opportunity for equipment redesign is limited.

RCM came out of an environment where, if the system was a problem, it could be redesigned. In most factories, buildings, and certainly fleets, the equipment is just a given of the equation. Some redesign can be done, depending on the capabilities of the organization. In most plants, redesign is very limited in scope.

Nine steps to PM Optimization

1. Task Compilation

Create a catalog of all tasks already performed by anyone who has contact with the machine. This catalog would include all current PM tasks (of course)

but also tasks done by machine operators, quality personnel, cleaners, calibration departments, safety inspectors, and others. Several departments touch the machine. It is essential to ask everyone who comes in contact with the machine what they do (specifically) and how often they do it.

2. Failure mode analysis

In RCM a great deal of thought and time goes into looking at the functions and the function failure engineering to determine all possible failure modes. In PMO failure mode analysis, the team works from the accumulated tasks back to the failure mode. In other words, failure modes without tasks are not considered at this time (they are considered later).

3. Rationalization and FMA (failure mode analysis) review

Rationalization is simple: put like causes together. This rule means that tasks addressed as the same cause are next to each other. Officially, if all the tasks from all the sources were loaded into a spreadsheet, then you can sort the spreadsheet by failure mode.

Equipment history is consulted to see if all failure modes are listed. The team reviews the engineering for the asset as well as the asset itself. They also use the experience of the old-timers. The team determines whether there are significant failures that are not covered by any task. Hidden failures are frequently in this last category.

4. Optional Functional analysis

Some analyses indicate an RCM type of functional analysis and loss of function. Such an analysis can be justified on highly complex equipment where the consequences of failure are severe. In these few examples, a sound understanding of function is essential to determine that all maintenance and operational issues are covered.

5. Consequences Evaluation

One of the break-throughs of RCM is its focus on consequences rather than on the failures themselves. The failures are looked at for consequences. The consequences divide themselves into two logical categories of Hidden and Evident. A hidden failure is the burning out of a warning light on an instrument panel (the failure is not evident since the light is normally out). A further analysis of the evident failure modes looks at the level of hazard and operational consequences. This aspect is analogous to the ESON factors in RCM.

6. Maintenance Policy Determination

This step is the core of PMO. Based on the consequences, certain decisions are made for each task. There is a series of questions for the PMO team to ask about each task.

The first determination concerns microeconomics. Is this individual task (labor and materials times frequency per year) worth the cost given the cost of the failure times the probability of the failure in that year?

Is there a better way to get to this failure mode? In some instances, introduction of quick condition-based monitoring would save overall time and money. The corollary is, would this task respond to simplification of the technology?

What tasks serve no purpose and can be eliminated. . Along with those considerations, which tasks can be set up at lesser or greater frequencies to increase effectiveness.

7. Grouping and review

This step is very practical in that it looks at the tasks that are left after duplicates and uneconomical tasks were eliminated and divides them up based on the facts of your operation. Questions that are answered include: Does operations perform all the daily tasks? Should the night shift be given accountability for this asset? Related to these aspects are tasks performed by different groups that cover overlapping areas. In the grouping step, the tasks are divided up so that each group gets the most appropriate tasks, given available skills, proximity to the asset, and scheduling factors.

8. Approval and Implementation

All parties have to be informed about what changed and why. All stakeholders are involved in this step. It is essential that the change be communicated to both maintenance and operations personnel and staff.

9. Living Program

Turning a PMO program into a living program requires time and patience. Less wasted PM will mean immediate freeing of resources. As these resources are reinvested to clean up the backlog, and the effective PM strategy takes hold, the number of breakdowns decreases.

RCM and PMO can rigorously recommend the right PM tasks and the right frequencies. Before we go into the details of effective PM, let's look at what PM is and what it does.

PM (Preventive Maintenance)

No series of initials or system has had as big an effect on the profession of maintenance management than the initials PM. The core concepts of PM are widely misunderstood and the methods of achieving reliability through PM are tricky and difficult. Yet, even with the difficulties when interviewed, 75-85% of the respondents would agree that they have a PM system in place. Most respondents would also agree that their system needed a little something, a little tuning up perhaps, to make it more effective. There is one thing without dispute. PM changes people's view of maintenance.

PM systems increase professionalism

One of the legacies we fight is the old concept of the grease monkey mechanic. Through the PM effort and other approaches, we need to increase our professionalism. In other repair fields such as computer repair and copy machine repair, professionalism is a job requirement. With the professionalism available to us, why do many PM systems fail to give us the advertised benefits?

PM systems often fail because PAST SINS wreak havoc on any maintenance department task force trying to change from a reactive fire-fighting operation to a proactive PM operation (particularly in a short time). Even after running for a few months, there are still so many emergencies that it seems you cannot make headway.

Why is this so? You face unfunded maintenance liabilities. The liabilities were generated when wealth was removed from the equipment without ade-

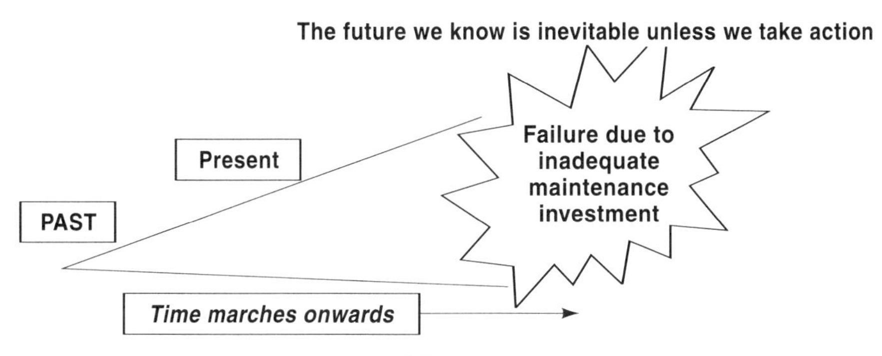

quate maintenance funds being invested back in to the equipment to keep it in top operating condition. Failures have a tail in the past where investments were not made.

Right at this very moment, hundreds or possibly thousands of unfunded maintenance liability items are in process in your plant. These deterioration processes have a tail in the past and a failure in the future. Proper PM will stop some of the deterioration, and inspection and PdM will detect the tail and estimate the time until the catastrophe. PM discipline will force a corrective action to repair, rebuild, or replace the asset.

There is a practical problem and a cultural problem. The only way through this jungle is to pay the piper, modernize, and rebuild yourself out of the woods. This situation is where the investment must be made, and is the practical part of the problem. Of course, any pitch to top management for a PM system must include a non-maintenance budget-line item for past sins, to bring the asset base up to the standard, (although it might properly be a capital expenditure).

The cultural problem is quite a bit more difficult to see, let alone to change. Your crews are probably addicted to the excitement and seeming freedom of a reactive environment. We call them excitement junkies! They will see their situation as an enjoyment of the variety of a reactive maintenance life. You will kill that with proper PM. The organization has to decide what kind of shop floor it really wants. The excitement junkies will still have some breakdowns, and as long as you don't have too many of them, everything will be fine (after a few years of consistency)

How do we see, interpret, and understand the future that we know is coming at us? How do we act is such a way as to fix these problems before they happen? The answer to both these questions is PM and PdM. Please note that, in much of the literature, PM and PdM are widely separated in thought and in action.

What is PM? PM is a series of tasks that either,

1. Extend the life of an asset.
   Example: Greasing a gearbox, cleaning electrical connections, or tightening a loose coupling

Or

2. Detect that an asset has had critical wear and is going to fail or break down.
   Example: A quarterly inspection shows a small leak from a pump seal. PM inspection detects the leak so that you can repair it before a catastrophic breakdown. Detection of critical wear, damage, or any other adverse condition, is considered inspection. PdM (discussed in detail later) is a specialized form of inspection using instruments.

Common tasks associated with PM are:

Type of task	Example
Inspection	Look for leaks in hydraulic system
Predictive maintenance	Scan all electrical connections with infrared
Cleaning	Remove debris from machine
Tightening	Tighten anchor bolts
Operate	Advance heat control on injection molding machine until heater activates
Adjustment	Adjust tension on drive belt
Take readings	Record readings of amperage
Lubrication	Add 2 drops of oil to stitcher
Scheduled replacement	Remove and replace pump every 5 years
Interview Operator	Ask operator how machine is operating
Analysis	Perform history analysis of a type of machine

Figure 31

These tasks are assembled into task lists. Each task is marked off when it is completed. Some tasks require readings or measurements. There should always be room on the bottom or side of the task list to note comments and readings. Actionable items should be highlighted to make it easy for the reader

These tasks should be directed at how the asset will fail. The rule is that the tasks should repair the unit's most dangerous, most expensive, or most likely failure modes (in that order).

Caveat: Even with the best PM systems there will still be failures and breakdowns. Your goal is to reduce the breakdowns to levels that are financially and operationally appropriate, consistent with a safe and secure environment. Through early detection, the breakdowns that occur will be of a reduced size and scale. Ideally, the breakdowns that are left will be converted into learning experiences to improve the delivery of maintenance service (and of course improve the PM system itself).

2nd Caveat (and this is a big one): Insure that whatever PM is performed you factor in iatrogenic failures. Iatrogenic is a fancy word that formally means: a symptom or illness brought on unintentionally by something that a doctor does or says. In our arena it means any breakdown or service due to an action of the mechanic or service that caused breakdown.

PM systems also encompass:

1. Record keeping systems to track PM, failures, and equipment utilization. Part of the job of the PM effort is creating baselines for other analysis activity.

2. All types of predictive activities. These activities include both human sense inspection and the use of instruments for taking measurements and readings. I Inspection of production for quality is included. PM incorporates the recording of all data for statistical and trend analysis.

3. Short or minor repairs. In this sense, short repairs refer to repairs that can be done completely and properly in a short time. Short repairs are a great boost to productivity because there is little or no lost time. Short repairs are to be written up for equipment history.

4. Writing up any conditions that require attention (conditions that will lead, or potentially lead, to a failure). Specifically this activity refers to writing-up corrective maintenance action items on work orders or work requests. Reports about machine condition are also included..

5. Scheduling, and actually doing these corrective repairs, written up by PM inspectors, within a reasonable timeframe (before they fail, for starters).

6. Keeping the PM process going and refining the task lists and task frequencies. One way to accomplish this work is using the frequency and severity of failures to refine the PM task list

7. Management of and investigation into trends uncovered by inspections.

8. Continual training and upgrading of inspector's skills, improvements to PM technology

Of course, the goal is not PM! The goal is the results from PM, higher reliability. PM is a way station to the ultimate goal of MI - Maintenance Improvement (high reliability without ongoing PM inputs). PM can be an expensive option because it requires constant inputs of labor, materials, and downtime. PM analysis should be designed to optimize PM and move toward MI

Appropriate Preventive Maintenance Strategy for the Three Equipment Life Cycles

All equipment and buildings require maintenance and they deteriorate in fairly characteristic ways. Good PM will slow this progression. Our maintenance and stores systems need to be sophisticated enough to detect what life cycle each piece of equipment is in, and react accordingly.

1. Start-up cycle.

The start-up cycle begins when you install the new equipment, and is characterized by infantile mortality. This mortality includes failures of materials, workmanship, and installation and/or operator training on new equipment. Frequently the costs are partially covered by equipment warranties (particularly for mobile equipment). By the nature of this life cycle there is a lack of historical data. Operators and maintenance people with long experience can anticipate some of the failures if they have experience with this sort of equipment. The failures are very hard to predict or plan for, and it is difficult to know which parts to stock.

Preliminary PM standards are developed in this cycle. Be careful not to weigh the failures in this cycle too heavily because, once the machine is debugged, understood, and is working properly, the early failures may never be repeated. This period could last from a day or less to several years on a complex system. A new punch press might take a few weeks to get through this cycle, and an automobile assembly line might take 12 months or more to completely shake down. Be vigilant in monitoring misapplications (the wrong machine for the job), inadequate engineering, and manufacturer deficiencies.

Although failures are unpredictable, there are actions (called countermeasures) that can be taken to reduce the problems and increase the probability that defects will be resolved in the initial life cycle and not be carried on to taint the next life cycle.

Countermeasures (designed to shepherd the asset through the first cycle and into cycle two)

- ❑ Operation and maintenance input into choosing machine
- ❑ Operation and maintenance inputs to machine design to insure maintainability
- ❑ Good vendor relations so you will be introduced to the engineers behind the scenes
- ❑ Good vendor relations so they will communicate problems other users have
- ❑ Participation in user groups, forums or discussion boards with other users of this type of machine.
- ❑ Communication with sister plants that have deployed this kind of equipment
- ❑ Enough time and resources to properly install
- ❑ Sufficient time to test run equipment
- ❑ Operator training and participation in start-up
- ❑ Latent defect analyses (run the machine over-speed, see what fails, and fix it)
- ❑ Formal procedures for start-up
- ❑ Maintenance person training in the equipment (and periodic retraining as project wears on)
- ❑ Rebuild or re-engineer to your own higher standard
- ❑ Maintenance person training in the field of commissioning or start-up engineering

2. Wealth cycle.

This cycle is where the organization makes money on the useful output of the machine, building, or other asset. This cycle is also called the 'use' cycle. The goal of PM is to keep the equipment in this cycle or detect when it might make the transition to the breakdown cycle. After detecting a problem with the machine or asset, the maintenance department will do everything possible to repair the problem and bring the asset back to life cycle two.

After proper start-up, the failures in this cycle should be minimal. Vigilance is necessary because with the machine running well there is a tendency to ignore PM. On the surface, PM yields very little at this point. Problems could crop up without notice. Without vigilance they could rapidly become catastrophic.

Attention to the PM task list is important in this life cycle. Using techniques from PM Optimization or other techniques, the PM task list can be pared down or bulked up, depending on your experiences.

A certified operator will minimize operator mistakes (some), sabotage, or damage, and provide early notice of material defects. Also PM would generate the need for Planned Component Replacement (PCR). The wealth cycle can last from several years to 100 years or more on certain types of equipment. The wealth cycle on a high speed press might be 5 years and might span 50 years for a low speed punch press in light service.

Countermeasures: Wealth cycle (designed to keep the asset in this cycle)

- ❑ PM system
- ❑ PM Optimization processes
- ❑ RCM processes
- ❑ Operator certification with periodic operator refresher courses
- ❑ Total Productive Maintenance by operators (TPM)
- ❑ Mechanic certification (as required by the airlines)
- ❑ Open communication with machine OEM or vendor
- ❑ Close watching and testing (such as oil analysis) during labor strife
- ❑ Audit maintenance procedures and check assumptions periodically
- ❑ Other vendors helping out with new products such as high tech lubricants
- ❑ Autonomous maintenance standards and quality audits
- ❑ Quality control charts to initiate maintenance service when control limits cannot be held.

3. Breakdown cycle.

This is the cycle that organizations find themselves in when they do not follow good PM practices. This cycle is characterized by wear-out failures, breakdowns, corrosion failures, fatigue, downtime, and general headaches. This environment is very exciting because you never know what is going to break, blow-out, smash up, or cause general mayhem. Some organizations manage life cycle three very well and make money by having extra machines, low quality requirements, and toleration for headaches.

Parts usage changes as you move more deeply into life cycle three. There is more reliance on the store room in this cycle because more parts are needed. The parts tend to be bigger, more expensive, and harder to get. The goal of most maintenance operations is to identify when an asset is slipping into life cycle three and fix the problem. Fixing the problem will result in the asset moving back to life cycle two.

Whenever you make a repair in life cycle three you revert to life cycle one until all problems in the repaired component has been shaken out.

Countermeasures: Breakdown cycle

- ❑ PM system to identify when bad things are going to happen
- ❑ Maintenance Improvement
- ❑ Re-engineering
- ❑ Reliability engineering
- ❑ Maintenance engineering
- ❑ Highly skilled maintenance workers
- ❑ Feedback failure history to PM task lists
- ❑ Great fire fighting capability
- ❑ Superior major repair capabilities and fully tooled shops
- ❑ Superior rebuilding capability
- ❑ Good relationships with major contractors that rebuild this kind of equipment

How to Justify Preventive MaintenanceExpenditures

To start with, the key is reducing breakdowns and the damage from breakdowns. We are vitally concerned with reducing the consequences to breakdowns. Consequences can be mitigated in a variety of ways. Eliminating the breakdown itself is one way. Without breakdown reduction, the resultant

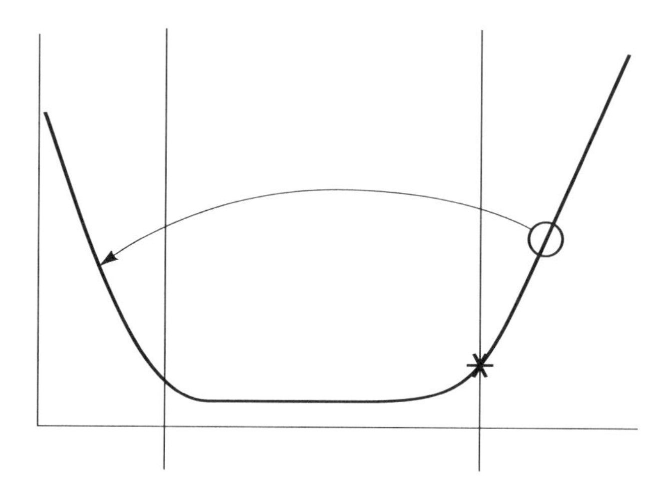

Repair in life cycle 3

downtime improvement, and the mitigation of other breakdown-related consequences, PM can be pretty hard to justify. The first chore in justification is to find out the true and complete cost and consequence of a typical breakdown.

What is the cost and impact of a Breakdown?

One way to sell PM is to discuss the effect of breakdowns on the whole operation. The cost of a breakdown is one element, and is greater than the downtime parts and labor. A whole variety of related costs might be included. Breakdowns and their consequences can be strong selling points when we want to develop a justification for improved PM. Some other consequences to consider:

Cost Justification if you don't have a PM program (will PM save us money?)

Calculate the costs of breakdowns. Be sure to add some of the costs from the list on the next page. Accumulate an average number of breakdowns per year and compare 70% of that cost with the cost of the inspections, adjustments, cleaning, bolting, lubrication, administration, short repairs, and corrective maintenance. We may assume that 70% of your breakdowns will be eliminated through an average quality PM system.

*(Number of breakdowns * average cost per breakdown * 70%) > Cost of PM system*

Some Consequences From a Breakdown

- ❑ Loss of goodwill
- ❑ Loss of customer
- ❑ Loss of customer trust
- ❑ Liability costs
- ❑ Safety consequences
- ❑ Increased workman's compensation costs
- ❑ OSHA, EPA Fines
- ❑ Damage to environment
- ❑ Operator (crew) idle time
- ❑ Lost production (less material costs)
- ❑ Loss of coverage of overhead
- ❑ Cost of rental equipment
- ❑ Cost of back up
- ❑ Extra travel time for mechanic
- ❑ Extra repair time due to adverse conditions
- ❑ Extra costs due to core damage
- ❑ Extra costs from overtime
- ❑ Extra damage of associated parts
- ❑ Extra costs of outside parts and labor
- ❑ Extra shipping costs
- ❑ Air freight
- ❑ Spoilage of product
- ❑ Discard of product in intermediate stages
- ❑ Disruption to job being interrupted
- ❑ Disruption to process and recovery
- ❑ Disruption to purchasing
- ❑ Disruption to storeroom
- ❑ Damage to morale

Your management needs to see the longer view on the nature of proper maintenance that only you can show. Sometimes managers hear from trade conventions or meetings that PM is hot this year and use this enthusiasm to help get a program approved or upgraded. Be conservative in your return on investment timing and liberal on the amount of funds it will take. Tangible returns may come late in the first year or second year (depending on the amount of deterioration and the speed of implementation).

The benefits possible from a PM program are real. Getting the benefit of the installation of a PM system requires a commitment to the elements of a successful system. To maximize the return of your investment in your equipment it must be kept in peak operating condition.

The PM Balance

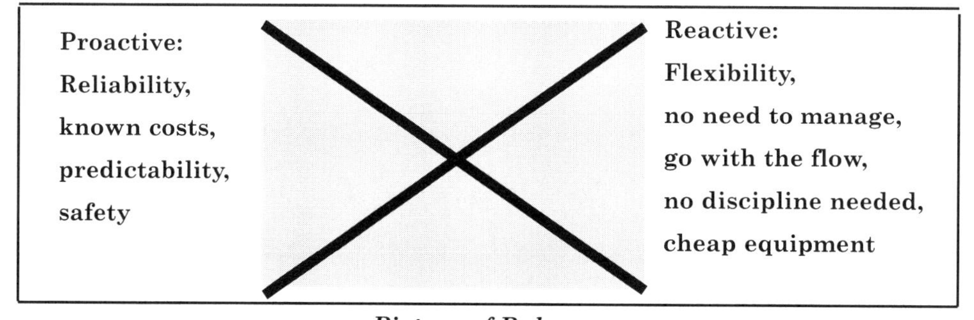

Picture of Balance

The PM approach is a long-term approach. Anything less than peak operating condition results in increased operating, maintenance, ownership, or downtime costs. These costs vary slowly. Low overall long-term costs of operation are the result of years of good maintenance policy.

Anyone can reduce maintenance costs for a few quarters by cutting back on PM inspections and the associated repairs. The temptation to do so is sometimes very great because the piper gets paid 1-2 years down the road.

Because of the temptation and the length of time to get a return on the investment most organizations have either no PM system or only a partial PM system. Some organizations inspect, lubricate, and adjust but don't feed back information on repairs to be scheduled unless there are clear and present dangers. Other organizations use fixed PM task lists with fixed frequencies without any review of failure, histories, or service.

Some other benefits of a PM system:

1. Your inspectors are your eyes and ears into the condition of your equipment. They provide a kind of human intel (like a spy on the ground). You can use their information on decisions to change your equipment make-up, change specifications, or increase availability.

2. Equipment has a deterioration curve (referred to as the P-F or Performance-Failure curve). Once over the threshold, failures increase rapidly and unpredictably. Working farther back on the curve adds predictability and reliability.

3. Early detection avoids core damage (allows you to rebuild the core rather than replacing it and having a reduced cost) and gives you more time to plan and secure parts and specialized tools.

4. Predictability shifts the maintenance workload from emergency fire fighting due to random failures, to a more orderly scheduled maintenance system.

5. The frequency of user-detected failures will decrease as inspectors catch more and more of the problems. Decreased user problems translate to increased satisfaction.

Pitching PM to Management:

The first step is to determine the cost of operating in your current mode. Be as complete as possible by adding some of the soft or intangible costs from the list on previous pages. Recruit people from marketing, accounting, quality, and safety, to apply numbers and (most importantly) the consequences from their perspective to each area.

When possible, include other departments such as production, accounting, or even marketing, to help prepare your arguments. Good maintenance effort affects every part of the plant, so every part of the plant has candidates to contribute to your discussions. Marketing is often a good choice because good maintenance will help them with the customers by ensuring delivery dates and maintaining quality.

The second step is to prove through rigorous modeling that savings or significant improvements to service will result from the proposed improvement.

In some applications the end customer will be the strongest voice for PM. All new vendors to General Motors are subject to a plant audit. One of the elements of the audit is the existence of a PM system (that seems to work). GM doesn't want to put its production in the hands of an organization that uses haphazard maintenance practices. If you were a giant customer like GM, wouldn't you require that your vendors have effective PM practices?

Maintenance costs may sometimes increase while overall costs decrease. The offset comes from decreased downtime, improved customer service, or another area.

Remember that we are in an extremely competitive battle for the organization's investment dollars. Less well publicized, we are also in a competition for management attention. Management has a short attention span. Although investments in maintenance can earn big returns they are not by themselves interesting or sexy to top management. That's where colleagues from other departments come in. They provide the sizzle.

Benefits of PM system: the stake holder's priorities.

Operations/Production	O	Purchasing	P
Maintenance Manager	M	Engineering	E
Top management	T	Accounting	A
Store room	S		

Benefit	Stake holders
1. Reduce the size and scale of repairs	M,O
2. Reduce downtime	T,O,E,M
3. Increase accountability for all cash spent	A
4. Reduce number of repairs	M,O
5. Increase equipment's useful life	A,O,E
6. Increase operator, maintenance mechanic, and public, safety	O,T
7. Increase consistency and quality of output	O
8. Reduce overtime	A,M
9. Increase equipment availability	O
10. Decrease potential exposure to liability	T,A
11. Reduce back-up and stand-by units	T,A
12. Insure all parts are used for authorized purposes	S,P
13. Increase control over parts and reduce inventory level	S,P,A
14. Decrease unit part cost	P
15. Improve information available for equipment specification	E
16. Lower overall maintenance costs through better use of labor and materials	M
17. Lower cost/unit (cost per ton of steel, cost per camshaft, cost per case of soda)	O,T
18. Improve identification of problem areas to know where to focus at tention.	M,O
19. Lower environmental fines	M, T, A
20. Reduced exposure to adverse publicity	T, O

We must sell our strong suits, which are cost avoidance, improved customer satisfaction, and reduced downtime. Other groups sell other benefits like the lawyer talking about lawsuits, or the environment specialist talking about fines and adverse publicity. Use the language (and issues) of your organization to sell a PM program. In every organization there are issues which are more important than any others.

Understanding PM (Preventive Maintenance)

PM task lists are the core of the program

The task list consists of the items to be done; the inspections, the adjustments, the lube route, cleaning, bolting, and readings and measurements. The two categories of PM activity are mandatory and discretionary. Sources of task lists in these two categories are:

Mandatory PM

- ❑ Laws (local, state, federal)
- ❑ Insurance companies
- ❑ Operating licenses
- ❑ Regulatory agencies (such as EPA, FAA, OSHA, DOD, DOT, DoE)
- ❑ Corporate standards
- ❑ OEM manufacturers (warranty)
- ❑ Your customer's requirements

There are several sources of Mandatory PM. In some instances the PM frequency and specific inspections are spelled out by an outside group such as a governmental agency, or insurance company. These entities can be very specific such as a State DOT-required vehicle inspection. The vehicle inspection tasks and frequencies are spelled out by regulations. In other examples, the mandatory PM is spelled out in general, with the specifics left up to the organization. In all instances, if you want to be in that business you have to adhere to the rules.

Discretionary PM

- ❑ OEM manufacturers (after warranty)
- ❑ Trade association recommendations

- ❑ Third party published shop manuals
- ❑ Skilled craftspeople experience
- ❑ History files with that equipment
- ❑ Consultants
- ❑ Equipment dealers
- ❑ Your customer's requirements (recommendations)
- ❑ Your own engineering department

Discretionary PM is primarily designed to increase reliability and reduce costs. You have to follow the recommended tasks and intervals to have equipment that is trouble free. The analysis is different because you have a choice; to follow the recommendations or not. The driver should be the consequences of failures. If the consequence is (real) safety or environmental problems, then the task must be done. If the consequence is economic (including downtime, labor, parts, product etc.) then an economic process must be followed to determine if the task should be done.

The Four types of task lists

Each type of task list has advantages and disadvantages. The four types of task lists can peacefully coexist on the same asset. Sometimes, the best choice is to mix the types so that the advantages of one type cancel out the disadvantages of the other.

Unit Based: This is the standard type of task list where you go down a list and complete it on one asset or unit before going on to the next unit. The mechanic would also be directed to correct the minor items with the tools and materials they normally carry (called short repairs).

TPM is generally a form of unit-based PM. In a TPM run factory, the operator is responsible for the unit PM. Usually a specific operator or team is responsible for the morning TPM cleaning, tightening, and lubrication.

Another variation of unit PM is called Gang PM where several people (a gang) converge on the same unit at the same time. This method is widely used in utilities, refineries, and other industries with large complex equipment and with histories of single craft skills. Gang PM is also used where PM requires a shutdown (designed to minimize downtime). The goal is to minimize the duration of the shutdown, so a gang is used.

Advantages:

- ❑ The mechanic gets to see the big picture because he or she is looking at the whole machine
- ❑ Parts can be put in kits and made available from the storeroom as a unit
- ❑ The PM can be engineered to reduce time and effort
- ❑ Work can be planned and scheduled
- ❑ Workers get into the mindset for the machine
- ❑ Mechanics learn the machine well and become expert
- ❑ Mechanics can discuss the machine with the operator, the supervisor, or management, from a knowledge-based point of view
- ❑ The mechanic (or TPM operator) has ownership
- ❑ Travel time is minimized (only requires one trip)
- ❑ Unit PM is a good place for training new people
- ❑ Work is easier to supervise than other methods because you know where the person should be

Disadvantages:

- ❑ High training requirements
- ❑ Still pretty boring
- ❑ Higher level mechanics are needed, even for the mundane part of the PM
- ❑ Short repairs can force you behind schedule
- ❑ If PM is not done, no one else looks at machine

String based: The second most popular type of PM strategy is termed string-based because the list is designed to hit one or a few items on many units in a string. The machines are strung together like beads on a necklace. Lube routes, gauge reading routes, rounds, and vibration routes, are examples of string PM. If the units are located near each other, it might be easier to look at one item on each unit. The inspectors' efficiency would be higher because each of them would be focused on one activity.

Most inspection-only PM's, are designed as strings, and almost all predictive maintenance is handled by various types of strings. One problem is that only a few computer systems (programs) support string PM (to support strings they must allow labor and material charges to be spread over several assets).

Advantages of strings:

- ❑ Low training requirements (in some applications, very low training requirements)
- ❑ Lower level mechanic is acceptable
- ❑ Jobs can be engineered with specific tools and exact parts
- ❑ Jobs can be engineered to reduce time and effort
- ❑ Jobs can be planned and scheduled
- ❑ Routes can be optimized to reduce travel
- ❑ Stockroom can pull parts for entire string at once
- ❑ Stockroom can deliver to pick-up points along the route
- ❑ Lends itself to just-in-time delivery of parts from vendors
- ❑ Standards for a string are easy to set
- ❑ Provides a good training ground for new people to learn about the plant
- ❑ Allows new people to get productive quickly

Disadvantages:

- ❑ Some loss of productivity, with extra travel time for several visits to the same machine
- ❑ May not see the big picture (the string person might ignore something wrong outside their string)
- ❑ Boring to do the same thing over and over
- ❑ No development of ownership
- ❑ Hard to supervise because the person is constantly moving
- ❑ If a mistake is made (such as wrong lube) it is quickly spread to all assets on the route.

Future Benefit: The future benefit type of task list takes advantage of closely-coupled processes, and is commonly considered in the chemical, petroleum, and other process-oriented industries.

For example, in a base-load, coal-fired, power plant in Utah, with three 500,000 MW units, there was an unscheduled outage due to a tube leak. While one crew was working on the tubes (with contractors who were specialists in that kind of work), another group jumped on PMs, NDT, corrective items waiting for shutdown, and other interruptive maintenance items. They maintained a work list in their CMMS that read status=waiting for shutdown.

The teams were given 48 hours for cool-down, repairs, and return to load. During that time, PM was accomplished on 14 or 15 major assets. The corrective jobs and PMs were pre-planned, so materials were on the shelf. Contracts were in place to rapidly expand the work force on short notice. In short, they

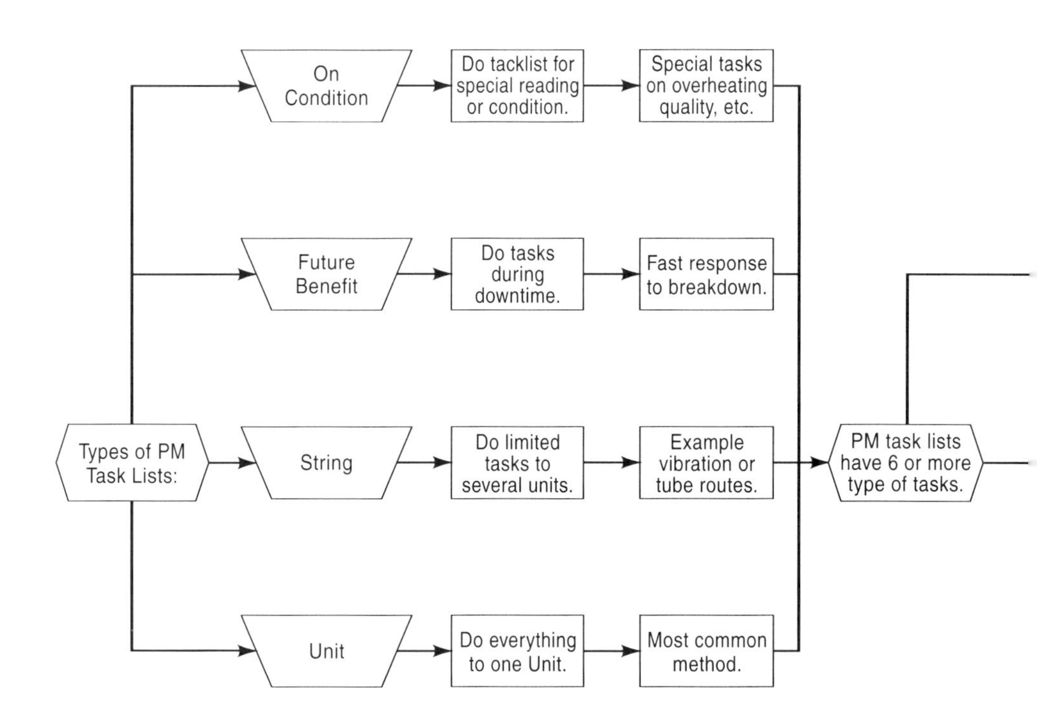

Figure 36 PM Tasks and Task List

were ready for an event (although they did not know what the event was going to be).

In future benefit PM, you PM the whole train of components whenever a breakdown or change-over idles one essential unit. This is opportunistic PM. It might not be a healthy steady diet, but if there is a breakdown or change-over, you might as well be ready to take advantage of the opportunity. It is

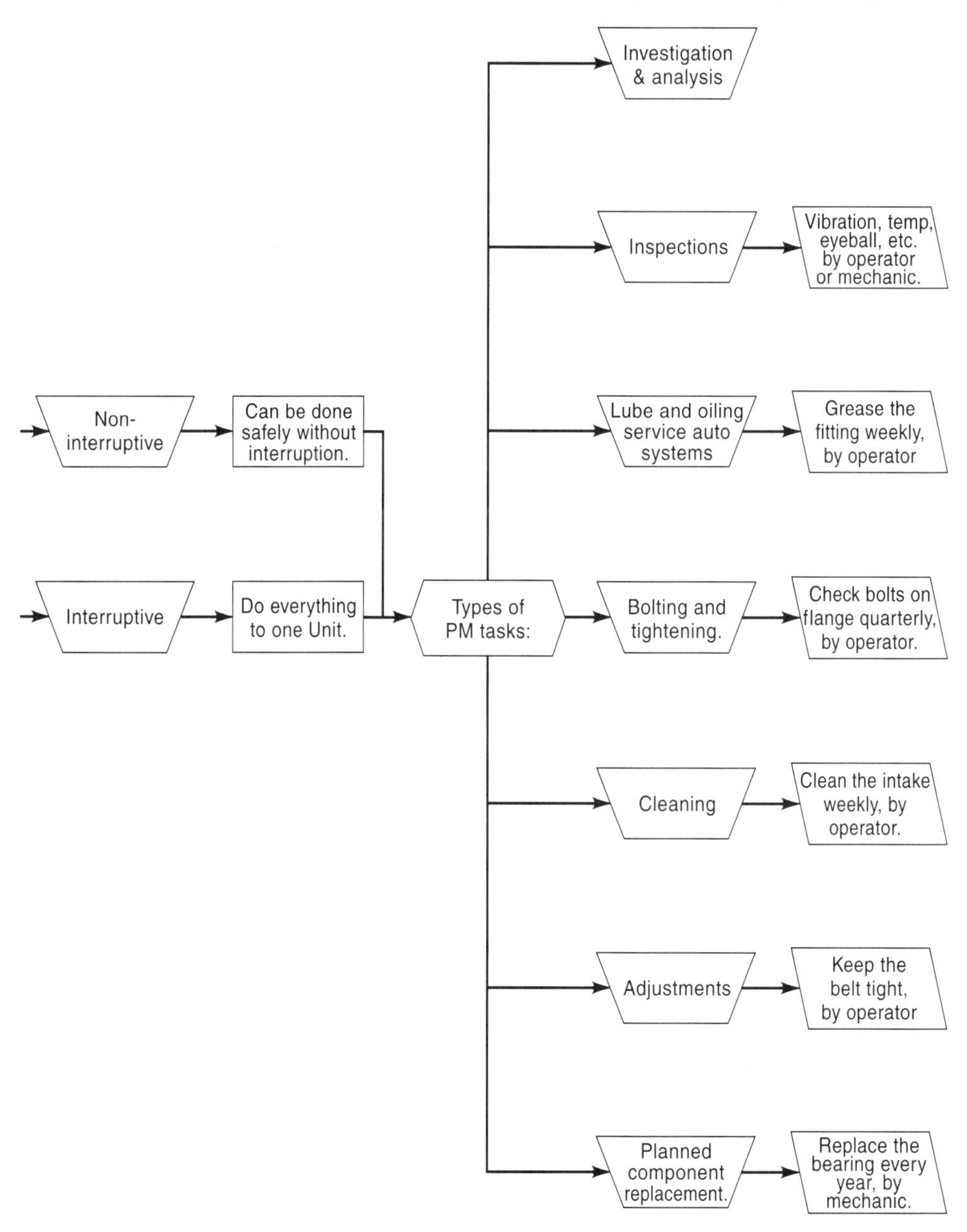

usually easier to extend downtime for an hour than it is to get a fresh hour for PM purposes.

The key to future benefit PM is to have the job plans complete and ready to go for any likely breakdown. The second key is to have adequate parts for the PM available in the storeroom all the time. The third key is to have a well-established call-out routine (either overtime or with a contractor) to get extra workers in a flash.

Advantages:

- ❑ Little or no additional downtime because the process train is already down
- ❑ Takes advantage of existing downtime to PM for a future benefit
- ❑ Can be made part of a larger PM effort to be used when the opportunity presents itself
- ❑ Can become a contest against time
- ❑ Easier to manage (intensive for a short time)
- ❑ Can be exciting
- ❑ Overtime opportunity

Disadvantages:

- ❑ Might not have enough people
- ❑ Disruptive to other jobs that are interrupted when the call comes in
- ❑ Disruptive to scheduled maintenance
- ❑ Disrupts people's lives
- ❑ Cannot predict when your next PM will be done, so you can plan but not schedule.
- ❑ Cannot pick your frequency using future benefit alone

Condition based PM: The condition-based PM service is based on some reading or measurement going beyond a predetermined limit. Some PdM inspections are condition-based (such as when electrical connections vary by more than 5°C). If a machine cannot hold a tolerance, a boiler pressure gets too high, or a low oil pressure light goes on, a specially-designed PM routine is initiated. This method is used with statistical process control to monitor and I ensure quality.

The automotive world has been accelerating along this road for years. Diagnosis nowadays consists of reading error codes off the automobile's brain. The processors aboard the vehicle collect codes for components that are wearing and need replacement or adjustment.

One great thing about condition-based maintenance is that the inspection can be conducted by computer and it can be conducted every second of every day. A condition-based PM might have a special task list for any specific condition.

Condition-based maintenance is a very big component of modern maintenance thought. Virtually all new equipment has microprocessors and can report on its own condition. We are stringing together even ancillary equipment on the factory network so conditions will be reported. As with the other methods, the condition based PM can be run alongside a unit or string-based method.

Advantages:

- ❑ High probability that some intervention is needed
- ❑ Less boring
- ❑ Involves the operator
- ❑ Brings maintenance closer to production
- ❑ Supports a quality program

Disadvantages:

- ❑ Warning might be too late to avoid breakdown
- ❑ Usually needs high skills
- ❑ May need expensive readers (the automotive version runs about $80,000)
- ❑ Can be planned but cannot be scheduled
- ❑ Many variations are not maintenance problems

Access to equipment

One of the most difficult issues of factory operation is access to equipment. Access problems fall into two categories; political and engineering.

Political access problems are problems that stem from political reality. The equipment is not in use 24 hours, 7 days. But it is in use whenever you want it for PM. The reason you are not given access might be because production control has not assigned time for the PM, or because the maintenance department is distrusted by production, etc. Some ideas for solving political access problems:

- ✓ Go back to the planning department to discuss requirements. Do not wait until you need the unit the next day. Sometimes the production schedule might be set weeks or months ahead of time. In fact, the discussion should happen at the beginning of the year and cover the entire year. PM activity should be consolidated, so that the production schedule needs to be interrupted as little as possible.
- ✓ Circulate PM success stories from your plant, sister plants, or from the trade press, to everyone in production management. Keep doing it until they believe you.
- ✓ Conduct a class in PM and breakdown, with examples of broken parts. Show how PM could have avoided the problem (of course you'll need to have been collecting broken parts and keeping them in the maintenance technical library display case for a while!)
- ✓ Use downtime reports now in circulation, and highlight downtime incidents that could have been avoided by PM effort.
- ✓ Use production reports in the same way.
- ✓ Most importantly, conduct yourself with integrity. Give equipment back when promised, show up when promised, and if there is a complication, communicate with everyone before, during, and after the event.

Engineering access problems

Engineering access problems are easy to spot. These access issues are with equipment that cannot be taken out of service because it is always in use. Full PM of this equipment might require a shutdown, or at the very least, a rental unit (such as a rental comressor). Consider transformers, environmental exhaust fans, and single compressors, etc. in this category

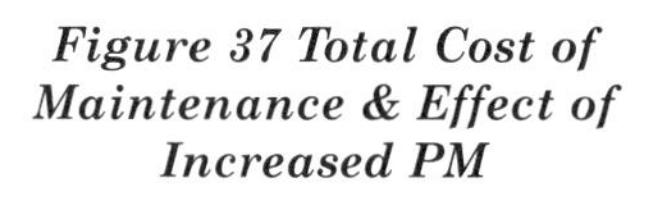

Figure 37 Total Cost of Maintenance & Effect of Increased PM

Work to minimumize total maintenance costs (around B)

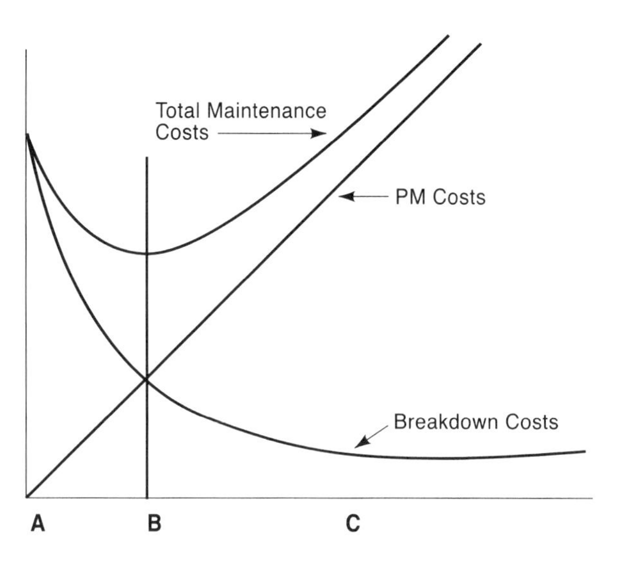

A partial antidote for both (engineering and political) equipment access problems might be in *non-interruptive maintenance.*

Interruptive/Non-interruptive: This category is a variation on the unit-based theme for machines that run 24 hours a day (or are running whenever you need to PM them). The unit-based list is divided into tasks that can be done safely without interrupting the equipment (readings, vibration analysis, adding oil, etc.) and tasks that require interruption. The tasks can be done at different times. The interruptive list may require half as much downtime as the original task list.

The next step is to re-engineer the specific machine so that almost all the tasks can be done safely without interruption.

Advantages: Same as above with reduced machine downtime
Disadvantages: Same as above, except it is slightly less productive since the machine requires at least two trips.

PM Frequency: How often do you perform the PM tasks?
PM Frequency and Its Effect on Breakdown

The chart on the proceeding page points out one of the difficulties in planning frequencies of inspection. The key is to minimize total maintenance costs. If the frequency (expense) of the inspections is too high, the total cost goes up. If PM costs are too low, then breakdowns are high and total costs are also high. Each operation has an optimum level of PM activity. We try to continually adjust the frequency and task list to move to the center of the curve (lowest overall cost).

A: The organization at point A has no PM system. All their maintenance demand comes from users, breakdowns, and routine work.

B: An organization in the B area has balanced its PM system to reflect the original failure rate. This level PM results in the lowest level overall cost for maintenance.

C: The C level organization has too much PM from strictly a cost point of view. Other factors such as government regulations, and high liability costs might drive excessive PM activity. . Airlines and nuclear power plants have both constraints and operate at very high PM levels.

PM frequency:

The first source for inspection frequency is the manufacturer's O&M (Operations and Maintenance) manual. Ignoring it might jeopardize your warranty. Keep in mind, when using the manufacturer's maintenance manuals, that some manuals are concerned with protection of the manufacturer, and limiting warranty losses (rather than minimizing your costs or exposure).

Following that manufacturer's guidelines may mean you will be over-inspecting and over-PM'ing the equipment.

One problem is that the manufacturer's assumptions for how the machine is being used might be different than your usage. For example, one manual recommended a monthly inspection for a machine. When the manufacturer was questioned, it came out that they had assumed single shift use. The factory used the machine around the clock and was getting excessive failures with recommended PM frequency. They might have had to go to a 10-day or weekly cycle to suit the manufacturer's recommendation when service was considered.

Certain mandatory inspections are driven by law (in the USA we have an alphabet of agencies that regulate different aspects of manufacturing, EPA, OSHA, and D.O.T. at both Federal and State levels. In some jurisdictions local agencies are added into the mix). You have a certain amount of flexibility in the timing of these inspections. Consider scheduling them when a mandatory PM is also due (such as changing your oil when the state inspection is due for your automobile). The rule is that while you have the unit under your control, you also perform any routine PM to improve efficiency. In fact, some CMMS can project PM requirements to a future date so that if the unit is down you are notified of any PMs due in the near future.

Your own history and experience are excellent guides because they include factors for the service that your equipment sees, the experience of your operators, and the level and quality of your maintenance effort.

The core of Failure

The original authors of the RCM methods thought long and hard about how machines fail. It is essential to understand failure to effectively design either tasks that will detect failure, or tasks that will extend life. In the P-F curve, failure is dissected.

One key is that these engineers determined that for each failure mode there is a precipitating event that starts the train of events. In this text the event is called the critical event (CE), and could be damage, contamination, heating, overloading, corrosion, even operations abuse.

The P-F (Performance-failure curve) describes the performance characteristics from the CE point to the ultimate total loss of performance that we generally call failure. Note that the slope of the curve (decay in performance) starts slowly and increases at an increasing rate. This sequence means that

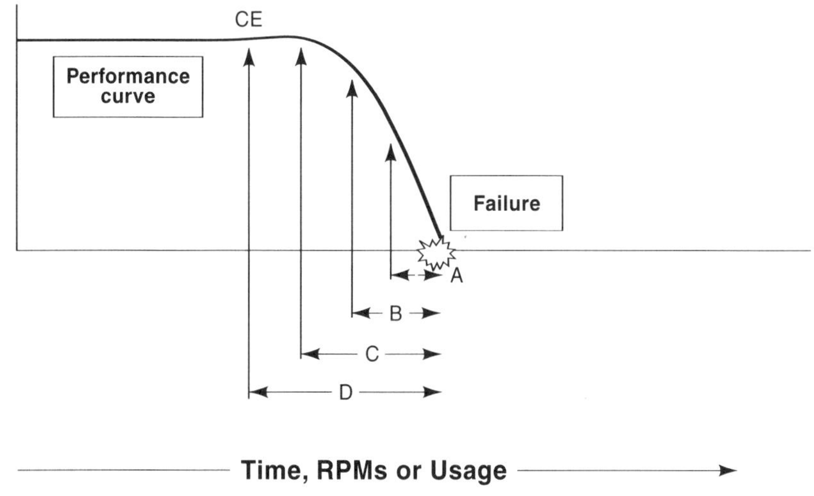

Figure 38 Performance Failure Curve

there is little or no discernable decay in performance in the beginning (near the CE). And, as time goes on, the performance decays rapidly.

The curve shows two factors that are essential for the proper determination of the task frequency. The first factor is how long is the interval between when an event happens and ultimate failure? This factor determines the frequency with which the the task must be performed for it to be effective. Any task must be done about twice as often as this interval. The second factor is what does Inspection and PdM buy you? In other words, what depth of task is necessary to achieve the reliability goal?

D- (The amount of time or use between the CE and the ultimate failure). Generally, this point cannot be detected with existing technology. It is a limit or a maximum theoretical time to failure.

C- is a point in the decay where very high tech PdM technology can detect either the decay in performance or the unfolding failure itself. The C point is the maximum practical time between inspection and failure. If PdM is used, the inspection has to occur sometime between points A and C. Generally, if you pick a frequency 1/2 (C-A), at least one inspection will fall between A and C. Using a PdM task gives you the greatest amount of time between detection and failure.

B-is a point where skilled technicians without technology can detect impending failure. No technology is needed at this point. This is the greatest interval without using PdM.

A-is a point where even unskilled workers, if told what to look for, will be able to detect the failure. Although anyone can detect the failure, there is not much time to intervene.

Whatever technique you use, based on the P-F curve, there are several ways to measure the interval

The PM inspection routines are designed to detect the failure modes unfolding and determine when failure will take place so that an effective intervention can be made. Since we cannot yet see the failure modes directly (point D), the goal is to find a measure that is easy to use and is more directly proportional to wear. Traditionally two measures were used:

Measures

For almost any measure to be effective, the PM parameter (such as cycles, days, etc.) must be driven from the unit level (unique parameter table for each unit), or from the class level (like units in like service have the same PM frequency).

Days: In the most common method, the PM system is driven from a calendar. (Example; every day, grease the main bearing, every 30 days replace the filter, etc.).

Advantages:

- Easiest to schedule
- Easiest to understand
- Best for equipment in regular use

Disadvantages:

- PM might not reflect how the unit fails
- Units might run different hours and require different PM cycles (example, one compressor might run 10 hours a week and another might run 100 hours)

Meter readings (such as hour meters): Example: change the belts after the compressor runs 5000 hours. Meter reading-initiated PM frequency is one of the most effective methods for equipment used irregularly.

Advantages:

- Relates well to wear
- Is usually easy to understand

Disadvantages:

- Extra step of collecting readings

- ❑ Hard to schedule in advance unless you can predict meter readings

Use: Second most common method. The PM system is initiated from usage such as; perform PM after every 50,000 cases of beverage, or overhaul the engine every 10,000 hours.

Advantages:

- ❑ Utilization numbers are commonly known (how many cases we shipped today)
- ❑ The parameter will be well understood
- ❑ Should be very proportional to wear
- ❑ Ties maintenance to production levels
- ❑ Not hard to schedule after production schedule is known

Disadvantages:

- ❑ CMMS information system might not accept this type of input
- ❑ Harder to calculate labor requirements per month or year (since you factor in the forcast)

Energy: The PM is initiated when the machine or system consumes a predetermined amount of electricity or fuel. The asset would have a meter or some other method of direct-reading of energy usage. This method is an excellent indirect measure of the wear situation inside the device and the overall utilization of the unit. In some instances you probably are already collecting some energy data for other reasons. Energy consumption includes the variability of rough service, operator abuse, extra heavy/dense/viscous raw materials, and component wear (increased friction). This method is used extensively on boilers, construction equipment, and stationary engines and turbines.

In mobile equipment the energy measure is gallons or fuel used (or poounds of propane).

Advantages:

- ❑ Very accurate measure of use in some equipment
- ❑ Raises consciousness about energy usage
- ❑ Can uncover other problems independent of maintenance

Disadvantages:

- ❑ Need to connect watt-meters or oil meters to all equipment to be monitored
- ❑ Hard to schedule ahead of time without a good history.

Consumables (example: Add-oil: The additions to hydraulic, lubricating, or motor oil are tracked.) When the add oil signal exceeds a predetermined parameter, the unit is put on the inspection list. This indicator is a direct measure of the situation inside the engine, hydraulic system, gear train, etc. Wear, and condition of seals, are directly related to lube consumption.

Advantages:

- ❑ Will alert you if there is a leak,

Disadvantages:

- ❑ Very specialized
- ❑ Very hard to schedule in advance
- ❑ Hard to collect accurate data

On condition measures (such as Quality: The incentive for PM here is generated by the inability of the asset to hold a tolerance, or have consistent output.) The call for PM could also be generated by an abnormal reading or measurement. For example, the low oil light on a generator might initiate a special PM.

Advantages:

- ❑ Responds well to customer needs

Disadvantages:

- ❑ Almost impossible to schedule
- ❑ Cause is frequently not in the maintenance domain
- ❑ Might be too late

Common tasks

The most commonly asked question is how to choose a task for inclusion in the task list. The general rule is to look at failure modes. How does the asset fail? If you have a history, review all the failures. Look for three types of failure situations: the modes that are most dangerous, most costly, and most likely. Any task list that sticks to these three is likely to be effective.

✓ Inspection: This approach is the low-tech, high-skill application of human senses including look at, feel, smell, hear, taste (only occasionally, such as in wine production).

- ✓ Predictive maintenance: This mode is inspection with some technological help. It could be vibration analysis, infrared scanning, or even megohm readings on a motor winding.
- ✓ Cleaning: A study in the Japanese Society for Plant engineering showed that 53% of all breakdowns in factories were caused by dirt, and bolting problems.
- ✓ Bolting/Tightening: Looseness is the second cause of machine break downs. Bolting includes looseness, missing fasteners, misapplied fasteners, and wrong fasteners.
- ✓ Operate: On some equipment that is used infrequently, observing the operation is the only way to ensure it will work when needed. Example: running an emergency generator for an hour each week.
- ✓ Adjustment: Many components fail because they are allowed to get out of adjustment, such as belts, limit switches, etc.
- ✓ Take readings: Many maintenance events follow unusual readings.
- ✓ Other events follow a slow decay in a key parameter that could have been detected by readings; such as boiler failure, air compressor failure, filter changes, etc.
- ✓ Lubrication: Basic PM includes lubrication, which is a critical item. Training is often slip-shod or non-existent. Investigate automatic lubricators, which present significant opportunities in this area.
- ✓ Scheduled replacement: Sometimes called planned scheduled replacement. Used effectively by airlines to produce ultra-high reliability. Excellent (but expensive) tool for mission-critical systems.
- ✓ Interview Operator: Ask questions and build a relationship of mutual respect. The operator is the closest to the action.
- ✓ Analysis: review the PM and repair history. Look for areas of possible improvement.

The one ratio always to consider when picking a task is:

What is the cost of the task per year related to the cost of the breakdown that is being avoided by this task?

Many people design task lists without considering the cost-benefit relationship.

Staffing the PM Effort

"A successful PM program is staffed with sufficient numbers of people whose analytical abilities far exceed those of the typical maintenance mechanic," (from August Kallmeyer, Maintenance Management). We want high-level people because they will be able to detect potentially damaging conditions before they actually damage the unit (point B in the P-F chart).

Your best mechanic is not necessarily your best PM inspector. The key to PM is to choose people who will do the jobs rather than people who say they will do the jobs. We also want people to apply their creativity by convincing others to change, improve, or modify the task list rather than showing their creativity by doing the task list differently each time.

PM is hard to verify (but it can be done), and usually boring for the people doing it.

Six Attributes of a great PM inspector:

1. Can work alone without close supervision. The inspector has to be reliable because it is hard to verify that the work was done.

2. PM inspectors must follow the list from first to last without improvisation. As mentioned above, we do not want people showing their creativity by doing the task list differently each time.

3. The PM person should be curious. The PM inspector is curious about the new advanced predictive maintenance technology and should be trained in the techniques of analysis and in the use of these modern inspection tools. The PM inspector should be curious about (and want to) review, the unit history and the class history in the CMMS (or manually) to see details of specific problems for that unit and for its class. We also want to choose the type of person who will fill out and complete the paperwork.

4. A PM inspector is pro-active in style. A mechanic is re-active in style. In other words, the inspector must be able to act on a prediction rather than react to a situation. He/she is primarily a diagnostician, not necessarily a `fixer.'

5. Some types of critical wear are subtle and difficult to detect. The more competent the inspector, the earlier the deficiency will be detected. Early detection of the problem will allow more time to plan, and order materials, and will help prevent core damage. Inspectors should use their experience and knowledge to convince others to change, improve, or modify the task list.

6. One of the things professionals look for is the right skill level for the task. With too much skill, the person will be bored, and with too little the person will be over their head (passed on Bill Thompson of Johnson Controls Services).

Scheduling

PM is a state of mind as well as a skill set. The state of mind has to be cultivated and maintained, which takes some effort. PM inspectors therefore, should be scheduled full time during the time they are assigned to PM. If full time is not possible, they should be assigned for a contiguous period such as from when they come in until lunchtime.

In any facility the PM inspection activity should represent 10- 20% of the hours of the whole crew. If everyone is rotated through the PM crew, be sure that they do PM for a whole day at a time.

In some places the PM crew is segregated from the rest of the maintenance crew. The crew might even be given additional duties and responsibilities. For example, in a large refinery the PM group is responsible for life safety inspection, and verification of quality in repairs. They consist of the top people, receive a pay premium, and have a different style of uniform.

Provide the PM inspector with the following to perform their tasks:

1. Task list with space for readings, reports, observations. The task list should include specs for the completion of the tasks and individualized drawings, if indicated. The task list should also include appropriate lockout information, confined space entry procedures, and permits.

2. Equipment manual or excerpt. Inspectors or operators should be encouraged to look through and familiarize themselves with the manuals.

3. Access to unit history files in the Maintenance Technical Library or the CMMS. In addition, every unit should be reviewed annually for trends and opportunities for maintenance improvements (this work should be a PM task called "analysis."

4. Standard tools and materials for short repairs. Operators should be given the exact tools needed for the PM or cleaning (10mm wrench, screw gun with 3/16 torx bit, etc.).

5. Any specialized tools or gauges needed to perform inspections

6. Standardized PM parts kits, lube, cleaning supplies

7. Log sheets to write-up short repairs (or PDA, laptop)

8. Forms to write-up longer jobs (or PDA, laptop)

Strategies to get PM done

Every plant has horror stories about machines failing right after a PM. The ongoing challenge of PM is to make sure the work is actually done and make sure the mechanic doesn't damage the equipment him/herself. This type of failure is called an iatrogenic failure (failure caused by servicing). We need to be vigilant about introducing more or new problems into any units we service. Failures of this kind should be investigated, and the equipment, procedure, or tooling modified to make iatrogenic failure unlikely (it is always lurking in the background).

Much thought, selling, coaching, and training are necessary to keep the PM crew involved and awake. This positive PM attitude needs the support of management, who should be listening to the inspectors and solving whatever problems they uncover (providing enough resources and downtime to repair or replace all corrective items).

1. In a pulp mill the supervisor would go through random PM's ahead of time and loosen bolts (non-safety related, of course), then check them afterwards. Sometimes he or she would make a game of it by telling the mechanic that 10 things were loosened.

2. An electrical supervisor taped `see me' cards inside panels. These cards could be traded for token gifts such as better parking spaces, lunch certificates, extra free time, etc.

3. One of the most interesting methods was the purchase of high-tech tools and training for the PM crew only. Using scanners, computers, and outside services, made the PM more interesting and more engaging.

4. PM can be reconfigured as a top level job by including some trouble shooting, job planning, sign-off when complete, safety inspections, and other extended role activities. Be sure to get rid of the nasty, grimy aspects, so that PM is looked at as a positive assignment.

5. Use the string PM for low-skill task list items. Have the PM inspectors supervise the strings

6. Be sure your plant rewards for uptime rather than for downtime. When asked if they were ever patted on the back or congratulated for machines that didn't break, only a few maintenance professionals could answer affirmatively. Most were rewarded for heroic failures that they worked hard to fix.

One of the best strategies to get PM done is Automatic Lubrication Equipment

There has been significant improvement in the reliability of auto lube systems. These systems can now be retro-fitted inexpensively to existing equipment on a one- or multiple-point basis. They provide a level of repeatability and reliability unmatched by most manual systems.

The biggest mistake is that organizations forget to add the lubrication equipment itself to the PM system.

STEPS to Install a PM (Preventive Maintenance) System

(partially adapted from The Complete Guide to Preventive and Predictive Maintenance by the author published by Industrial Press)

1. Create PM task force. This group includes craftspeople (plus the shop steward in union shops), a staff representative, data processing representative, and engineer(s). In some operations, a representative from operations is essential. This task force thrives where there is also a supporter from management (but the supporter doesn't have to be a member of the team). Keep in mind that PM has four dimensions (engineering, economic, psychological, and management). Members of the task force should have expertise in one or more of these four areas and all four areas should be covered.

2. Decide on the goals of the task force and of the whole program. Set objectives. Begin to design the training program. Everyone on the task force must become an expert in PM (most people should learn much more than they know now). Do not start to design the program until the task force personnel are well trained in PM.

3. Pick a catchy name for the effort like PIE (profit improvement effort), DEEP (downtime elimination and education program, QIP (quality improvement program). Stay away from PM unless you can establish that PM does not have negative connotations for a majority of the stakeholders of maintenance improvement.

4. At some time early in the process, the task force should decide that PM is the appropriate strategy for the organization at this time. Prepare a complete economic Return-on-Investment (ROI) case study. Macro-analysis will determine whether the PM strategy is best for your organization. This economic analysis is essential to enroll top management in the process. The process is essential to get top management commitment, and to get funds for the next steps. The case study should show the costs of the current operation, costs of the proposed operation, and the costs to get from point A to Point B.

5. Get training in computers for members of the task force if they are not computer literate. Get them access to computers and any relevant organizational level networks or systems. At a minimum the task force should be able to use word processors, spreadsheets, e-mail, and presentation software. Much of the work of the task force can be shared by e-mail and can be designed in a Lotus Notes type of environment. Intranet

users can start a web site to inform members of the task force, and eventually the rest of the organization. Of course, training in PM also is well under way.

6. Part of the PM training is maintenance management training. Get generalized maintenance management training for the entire task force. This training will save time and effort by laying the groundwork for them to share a language and create a new vision of maintenance. There are many good teachers in every part of the world, and this training is money well spent. Some organizations build expertise by using a variety of the leading trainers to build a unique vision for PM for their organization.

7. Identify the maintenance stakeholders (that is, anyone impacted by how maintenance is conducted). Analyze the needs and concerns of the maintenance stakeholders. Use questionnaires, interviews, and common sense to determine each stakeholder groups' stake in the outcome. Look at each group and see how they contribute to the success of the organization. Tie in the plan to that outcome. This exercise is like showing how a reduction in breakdowns is demonstrated to reduce lost-day accidents when presentations are made to the risk management or safety people. Include production, administration, accounting, office workers, housekeeping, legal, risk management, warehousing, distribution, clients, etc. At least look at how your proposed changes will benefit each group. Consider drafting an impact statement for every group your change will affect.

8. Once the stakeholders are identified it is time to design and deliver the important "dog and pony show." This show is a (Power Point) slide presentation given to various stakeholder groups, that describes the program, lists the steps, and builds enthusiasm around the outcome. It should train the stakeholder in PM as it impacts them. Different shows can be developed for major stakeholders. Several (all) members of the task force should participate in the presentation. Enlist other groups to talk about their areas of specialty. Examples include talks to production about the consequences of downtime, to accounting about the impacts of costs, to marketing on the importance of the plan to smooth out delivery problems, and so on. An especially effective tactic is to be sure that a member of the stakeholder group to whom you are speaking gives a part of the presentation. It is important to realize that, no matter how good your PM plan is, and no matter how bad the existing situation is, stakeholders will think they have something to lose through the change and nothing to gain. until they are convinced.

9. Once buy-in has occurred, the core work of the project can begin. Design Key Performance Indicators (KPIs) for the project. There should

be several simple measures that will show the team (and other interested parties) how the project is going. With all KPIs, an up-to-date display of progress (or lack of it) is a powerful way to keep the project going. A display in an easy to understand format like gauges on the dashboard of a car (Joel Leonard's idea), a dial, or even a rocket going toward a goal, will shows the data in an easy to understand format. The key is to keep it up to date and accurate. Some KPIs to consider:

A. The first step is building the master files so use a simple thermometer showing the percentage of assets entered and audited into the system. If there are other master files, perhaps a measure of all of them, or critical ones only.

B. Once the files are built, everyone must be trained. The percentage of training completed could be another indicator. A way to express this might be an estimate of all training to be done, divided by training completed.

C. As areas are covered and work orders start to be issued, track hours reported by the system compared with payroll hours for the same group.

D. As PM tickets get issued, track PMs issued and completed each week.

10. Make inventories and tag all equipment to be considered for PM. Compile and review your list of equipment. If you have an effective CMMS then this step should already have been completed. If so, audit the master lists of equipment. Pass the list out to operations for verification. This list is a starting point for the PM program. Inquire if lists exist in plant engineering or accounting.

11. Select the system (CMMS is the usual choice) to store information about equipment, select forms for PM-generated MWO and Check-off sheets. Again, if you already use a CMMS then the choice is complete. The challenge is to build the task forces' expertise in the PM module for your particular system. All the CMMS are slightly different. These differences might seem trivial on the surface but they could make the job much harder. Expertise with the specifics of your system is essential. If there is one, try to attend a user meeting for the CMMS.

12. Design first drafts of the KPIs to be used to evaluate the new PM system's performance. At different stages in the projects, different KPIs are needed to move the project along. These measures will be revised as the process goes on. Some of the KPIs designed in paragraph 9 above can be morphed into the new drafts with little extra effort.

13. Take a complete look at your business process. Chart the steps necessary to get PM done. Consider changing the business processes to speed them up and reduce the time needed. As the new process is designed, begin to draft standard operating procedures (SOP) for the PM system. These procedures will also be revised many times over the first year. After your business process is charted and understood, see if

new technology will enable you to use a new way to run the process.

14. Ask your task force members or shop personnel to complete data entry, or prepare equipment record cards (if not already complete from the CMMS installation). Rotate the data entry job so that many (everyone?) in the department has experience collecting, adding, and auditing data. Widely-held experience in correcting mistakes in the database is essential before you go on line. **It is essential to build a critical mass of expertise in the system before you go live.** If this is a CMMS installation, there are two levels to the effort. One level is to collect a complete list of all equipment. The second level is to collect all the nameplate data and add that to the files. These two different tasks can be done sequentially, or at the same time. If the CMMS is already operational, you can start to build the PM module (but the rules above still apply). Enter the task lists and frequencies, and relate them to individual equipment. Enter details of parts data, parts kits, tools required, lock-out/tag-out steps, PPE, PM task steps, and other planning data to make the PM tickets instantly useful.

15. Be sure to replace hours invested in the system. Bootstrapping the PM system will hamper the effort and make it take longer. Consider using contractors and some overtime to replace the hours lost on the floor by the people doing the data entry.

16. Fight the tendency to use the CMMS vendor to build the details of the PM system. Vendor employees can be directly involved (if you feel you need the expertise) but only as advisors. Ideally, when needed, vendors should be hired as advisors, auditors, cheerleaders, and councilors, but not on the playing field as team members.

17. Another essential step is the ongoing daily audit of all task list and support data typed into the system. Have someone who is highly skilled review all data going into the system.

18. Select people to be inspectors. Allow them to have input into the next steps. Consider using inspectors to help set up details of tasks in the system. This group will be handed the system after the task force has been dissolved.

19. Arrange for key personnel to be trained in RCM (reliability centered maintenance) or PMO (PM Optimization) and failure analysis. This training will help them and the program immeasurably. The training will show them how to root out useless tasks and include important tasks on hidden functions.

20. Determine which units will be under PM and which units will be left to break down (BNF or bust and fix). Remember that there is a real cost

associated with including any item in the PM program. If, for example, you spend time on PM's for inappropriate equipment, you will not have time for the essential equipment. Calculate the cost to include an item in a PM Program from:

Cost of Inclusion = Cost per PM * Number of PMs per year.

To decide which units to include in the PM system, apply the following considerations to each item:

A. Would failure endanger the health or safety of employees, the public, or the environment?
B. Is the inspection required by law, insurance companies, or your own risk managers?
C. Is the equipment critical?
D. Would failure stop production, distribution of products, or complete use of the facility?
E. Is it the link between two critical processes?
F. Is it a necessary sensor, measuring device, or safety protection component?
G. Is the equipment one of a kind?
H. Is the capital investment high?
I. Is spare equipment available?
J. Can the load be shifted easily to other units or work groups?
K. Does the normal life expectancy of the equipment without PM exceed the oper ating needs? If this is true, PM may be a waste of money.
L. Is the cost of PM greater than the costs of breakdown and downtime? Is the cost to get to (to view or to measure) the critical parts, prohibitively expensive?
M. Is the equipment in such bad shape that PM wouldn't help? Would it pay to retire or rebuild the equipment instead of PM?

21. Once it has been decided that an asset (machine, unit etc.) is to be included on the PM system (the macro analysis has been completed) the microanalysis begins. A look at the failure history, collected information from the OEM, and the accumulated experience of your maintenance and operations team are to be focused on that machine.

22. Schedule modernization on units requiring it. Investigate the possibility of retiring bad units. A bad unit left on the system will demoralize the most dedicated inspectors.

23. Select the PM clock you will use (days, utilization, energy, add-oil). A clock is designed to indicate wear on an asset. Clock times on items in regular use or subject to weather are usually expressed in days. An irregularly used asset might be better tracked by usage hours or output tons of steel, cases of cola, etc. Some units, such as construction equip-

ment, are best tracked by gallons of diesel fuel consumed because hour meters are frequently broken.

24. Decide what Predictive Maintenance technology you will incorporate. Train inspectors in techniques. Even better, provide the information and a budget to the task force and let them pick the technology. Most equipment should be rented before buying. Inexpensive training is available from most vendors and distributors.

25. Set-up task lists for different levels of PM and different classes of equipment. Factor in your specific operating conditions, skill levels, operators experience, etc. Consider all the strategy, including unit-based, string, route maintenance, and future benefit, as well as non-interruptive /interruptive. Consider what strategy to use to schedule PMs, and what to do if the date slips. Be sure to include a review of the failure history when designing the task list. It is great to design for possible failure modes but it is essential to design for the failures that occur. .

26. Start a program of public relations. Sift through your data and find statistics that indicate success stories. Find out those stories and write them up as powerful narratives. Publicize your successes. It is okay to publish stories from your industry or from other industries where similar equipment is involved. One idea is to collect PM stories or maintenance catastrophes (that you can show would not have happened with PM) from various sources, and circulate details of a different one every month.

27. Document all PM tasks. Categorize the PM tasks by source (recommended by Ron Moore, of RM Group). Categories might include regulatory, calibration, manufacturer's warranty, experience, insurance company, quality, etc. Documentation will be a great aid when you look back to see which tasks to eliminate or change.

28. Provide the PM inspector with the following items to perform the tasks:

A. Task list (usually printed on a work order) with space for readings, reports, observations
B. Drawings, performance specifications, pictures where appropriate
C. Access to unit history files, trouble reports.
D. Equipment manual or excerpt
E. Standard tools and materials for short repairs.
F. Consider cart designed for the PM's and common short repairs
G. Any specialized tools or gauges to perform inspection
H. Standardized PM parts kits

I. Forms for writing-up longer jobs to be submitted to the maintenance dispatcher.

J. Log type sheets to log short repairs or (if your system will allow it) short repairs are added to the bottom of the PM sheet and entered into the system. It is important to capture short repairs in the CMMS (if possible).

29. Assign work estimates and job steps to the task lists for planning purposes. Observe some jobs to get an idea of timing. Define tools and parts necessary for each PM. Have the lists verified by the mechanics that actually do the PMs.

30. Let some mechanics time themselves and challenge them to re-engineer the tasks to cut PM time. Remember that time spent on PM does not itself add value to your process. The goal always is to minimize PM time while getting the task done correctly.

31. Engineer all the tasks. Challenge yourself to simplify, speed-up, eliminate, and combine tasks. Improve tooling and ergonomics of each task. Always look toward enhancing the worker's ability to do short repairs after the PM is complete.

32. Determine frequencies for the task lists based on the clocks chosen. Select parameters for the different task lists.

33. Implement system, load schedule, and balance hours. Extend schedule for 52 weeks to coordinate with production control. Slots should be carved into the production schedule for the whole year.

34. Balance PM to actual crew availability. Schedule December and August lightly or not at all. Allow catch-up weeks throughout the year.

35. As PMs get generated and completed, be sure to monitor KPIs such as PMs generated to PMs completed, as PM schedule compliance. Formally and informally, ask if the PM job plans were accurate and fix any problems.

36. Plan to have a periodic meeting with the task force (as it is now constituted) to evaluate the on-going use of the system. Look at uptime increases, reduced size and scale of repairs, and trends in the KPIs.

TPM Total Productive Maintenance

There is a revolution going on, on the factory floors of selected organizations. The ideas of TPM are to make the operator an equal partner in the maintenance effort. This idea, imported from Japan, has taken root in factories, refineries, mills, and power plants throughout North America. It succeeds because it forces us to realize that we have to use more and more of the capabilities of every employee (and vendors too!) to remain competitive. The group of operators is traditionally viewed as underutilized in most factories.

The machine operator is the key player in a TPM environment. Many of the losses are under the control of the operator, involve the operator, or happen while the operator is near the machine. There is less reliance on the maintenance department for basic maintenance (but more for coaching, training and mentoring). Control and responsibility are passed to the operators.

There is complete focus on the losses from production. The losses can come from six areas. Rigorous data collection followed by analysis is necessary to identify losses in each of the six areas.

TPM can be summarized as attention to and elimination of the six losses of production:

DOWNTIME:

1. Equipment failure from breakdowns. This is the biggest element that is directly the responsibility of maintenance. With TPM, first line maintenance activity is transferred to operations. Proper design insures reductions in breakdown-related downtime.

2. Setup and adjustment. The stated goal is single digit minute set-up times. This timing allows up to 9 minutes for set-up. Adjustments are simplified or eliminated from the system. Overall re-engineering to reduce these exposures is expected.

SPEED LOSSES:

3. Idling and minor stoppages due to abnormal operation of sensors,

blockage of work on chutes, etc. These slow-downs are tracked and analyzed to see what is really happening. Analysis of root causes and of processes is continued until the system no longer has losses in these areas.

4. Reduced speed due to discrepancies between design and actual speeds. Design speeds are reviewed, and actual speeds are observed. If unfavorable, in comparison, initiates a design and engineering review.

DEFECTS:

5. Process defects due to scrap and quality defects to be repaired. Quality problems are not tolerated. Deep analysis is undertaken until these losses approach zero.

6. Reduced yield from start-up to stable production. The production process is tracked and watched for start-up problems. Any start-up problems uncovered are fixed permanently. Stable production should follow start-up very closely.

History of production

In many ways, TPM is a return to a pre-1920's model of maintenance. Before the 20's, machine operators were skilled craftsmen. They were expected to repair their own machines. Some served lengthy apprenticeships. As mass production took over, lower-skilled operators were recruited and the production jobs became more menial. Many of these newly-minted operators were immigrants or just off the farm. Their greatest asset was the low wages they would accept and the long hours they worked.

The ability to fix one's own machine quickly was gone. Training to improve one's skills was not common. Soon this group, as well as management, forgot that these people had capabilities far exceeding their needs as operators. A tradition settled in of operators being only button pushers. The maintenance department, as we know it, developed at that early time by necessity, with specialists in repairs and maintenance.

Total Productive Maintenance (TPM) has four elements

TPM is one of the most effective methods of improving the delivery of maintenance service while increasing the effectiveness of the equipment. Although in its entirety TPM doesn't apply in many situations, some aspects apply to all maintenance situations. TPM is the maintenance department's answer to the empowerment, job enrichment, and total quality programs on the production floor. The great advantage is that TPM can be incorporated into and can greatly enhance these programs. To begin with "The dual goal for TPM is zero defects and zero breakdowns." To achieve this goal, TPM has four elements:

1. Maximize overall equipment effectiveness. TPM has a very strict definition of effectiveness. One of the tenets of TPM is that sloppy readings of effectiveness can cover up opportunities for production improvement.

2. TPM establishes a shared system of PM for the equipment's complete life (it takes into account the life cycle of the equipment). PM should be modifiable based on the life stage of the equipment. Without this modification ability, PM tasks might not reflect the failure modes of equipment in that condition. The shared PM divides the tasks up between production and maintenance (In a manner that is very similar to task assignment in PMO).

3. TPM must be implemented by all departments including maintenance, engineering, tool/die design, and operations, etc. Like many other programs of this type, TPM is not really a maintenance program but rather a partnership between maintenance and production. The partnership will affect all the other stake holders of maintenance. Their involvement is necessary for TPM to thrive. Thus, every employee must be involved in TPM from the workers on the floor to the president.

4. TPM is based on the promotion of PM as a motivational technique through autonomous maintenance groups (operators have greater involvement and say about equipment). TPM works only because the operators begin to own the equipment. As ownership spreads, autonomous maintenance becomes a reality.

Why the results from TPM are urgent today

A new situation has developed in the way we look at organizations. Throughout the early 1990's, organizations slimmed ranks, reduced overheads, and optimized processes. At the same time, we increased the complexity and speed of the equipment and our reliance on computers, PLC's, and sophisticated controllers. We were faced with smaller crew sizes and basic maintenance demands were going unmet.

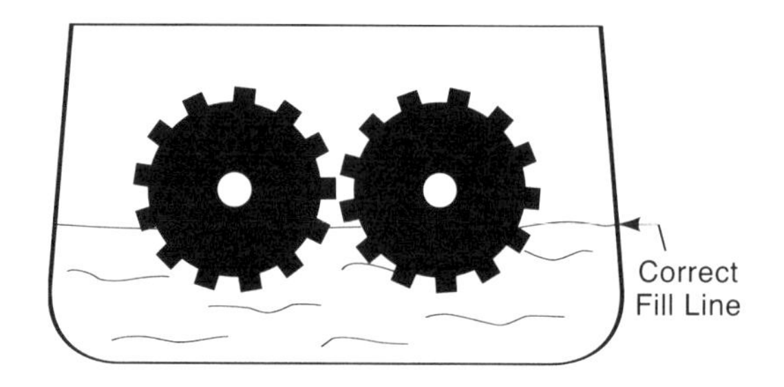

Figure 39 TPM typical specification for maintenance work.

TPM recruits the operators into the maintenance function to handle basic maintenance tasks and be the champion of their machine's health. TPM returns to pre-1920 roots by re-involving the operator in maintenance activity and decisions.

The maintenance department becomes an advisory group to help with training, setting standards, doing major repairs, troubleshooting, and consulting on maintenance improvement ideas. Under TPM, maintenance becomes very closely aligned with production. For TPM to work, maintenance knowledge must be disseminated throughout the production hierarchy. The old philosophy of `produce at all costs, damn the torpedoes - full speed ahead' will fall flat on its face with TPM. The operators must have complete, top level support throughout all phases of the transition and thereafter.

TPM uses the operators, in autonomous groups to perform all the routine maintenance including cleaning, bolting, routine adjustments, lubrication, taking readings, start-up/shut down, and other periodic activities. The maintenance department becomes specialists in major maintenance, major problems, problems that span several work areas, and trainers.

Installation of TPM

Three basic requirements prepare the soil for the transplantation of TPM concepts and attitudes:

Many of the steps for a successful PM installation also apply to a TPM program except that the group is larger and more widely distributed throughout the plant.

Motivation: The whole staff and all the workers need to be open to a change of attitude toward waste. This change will take place over long periods of time as organizations present, train, and start to change toward TPM. TPM has to be viewed as interesting, challenging, and fun for maximum acceptance.

Competency: Certain skills are necessary before TPM can succeed. Training operators in PM, and design engineers, and mechanics in root failure analysis, will eliminate waste and losses.

Environment: The thrust for improvement must be supported by the top managers in the organization. The top people must understand the need for and the implementation of TPM.

Steps

In the Japanese form of TPM the operator goes through seven steps to reach full autonomous maintenance authority.

1. Initial cleaning, review of entire machine, tightening
Complete cleaning of machine. Repair any deficiencies that become apparent during the complete cleaning. Part of this process is tightening of all fasteners to specifications, and review of the entire machine operation.

2. Maintenance prevention
Reduce time to perform cleaning. Remove sources of contamination. Make the machine easier to service (lubricate, tighten, clean, adjust).

3. Establish consistent standards
Specify all tasks and frequencies (daily, weekly, every 1000 pieces, etc.). Set standards for tasks (how clean, what to use to clean, how much and what type of lubricant). Autonomous group prepares documentation.

4. Inspection
Initial inspection follows manufacturers' manuals, engineering recommendations, and equipment history (what has failed). Group is taught how to correct minor defects.

5. Autonomous Inspection
Inspection is turned over to group. Check sheets are utilized for all inspections. Minor repairs are completed. Maintenance is involved only in major problems that involve specialized knowledge, skills, or contacts.

6. Organization to support ongoing TPM
Systematize the autonomous maintenance activity. Align the organization to support TPM. Use the TPM productivity reports to run the plant. Develop standards for all activity.

7. Full functioning TPM
Track the results of the effort and give ongoing recognition to progress. Monitor failure frequency and look for additional improvements. Spend more time on improvements that reduce maintenance effort while increasing equipment availability.

In Western implementations the steps are somewhat different

1. Introduction to TPM
Conduct presentations to all levels of management and workers before roll out. Initial training of what TPM is (and what it is not) and why it is important to the operator. Analysis of what exact role maintenance will play in the TPM teams. Build TPM into the reward and review structure of everyone in production from the operator up to the production manager. TPM goals are built in to the incentive system. Decisions about TPM data collection and whether data flow will be under CMMS.

Budgets for maintenance improvement are approved, so that some funds are available when improvements are discovered.

2. Design and implement training program
Begin the basic training of operators in all the skill sets indicated by a review of the equipment. Competence grids created for each machine with training and testing designed. Stage three should begin in the middle of this stage so that the operators see where the training is leading.

3. Initial cleaning, inspection, and review of entire machine
Complete cleaning of machine by the TPM operator team under the coaching of a qualified (and TPM trained) maintenance facilitator. Repair any deficiencies that become apparent during the complete cleaning. Part of this process is the tightening of all fasteners to specifications and review of the entire machine operation. Initial inspection follows manufacturer's manuals, engineering recommendations, and equipment history (what has failed in the past). The TPM group is taught how to correct minor defects.

3. Establish consistent standards
Specify all tasks and frequencies (daily, weekly, every 1000 pieces, etc.). TPM tasking is explicit. Set specific standards for tasks (how clean, what to use to clean, how much and what type of lubricant). TPM teams prepare documentation and determine what additional training is necessary so that each TPM worker can complete the competence grid for the equipment for which they are responsible.

4. Maintenance prevention
TPM teams look into what is driving the need for maintenance, and initiate projects to reduce maintenance requirements. These projects are already funded by budget additions at the outset of TPM. Reduce time to perform cleaning. Remove sources of contamination. Make the machine easier to service (lubricate, tighten, clean, adjust).

5. Autonomous Inspection
Inspection, Tightening, Cleaning, and Lubrication are turned over to the TPM group. Check sheets are utilized for all inspections. Minor repairs are completed. Maintenance is only involved in major problems that involve specialized knowledge, skills, or contacts. Maintenance may also be used for QA purposes.

6. Full functioning TPM
Systematize the autonomous maintenance activity. Use the TPM productivity reports to run the plant. Track the results of the effort and give ongoing recognition to progress. Monitor failure frequency and look for additional improvements. Spend more time on improvements that reduce maintenance effort while increasing equipment availability.

Measuring equipment effectiveness is an essential part of TPM.

Many organizations do not or cannot capture accurate information about run times, slow downs, minor stoppages, and defects. TPM relies on good record keeping in the six areas of loss. The following chart case study show the six big loses and a calculation of effectiveness for a manufacturing machine with a 3 second cycle time. The chart on the following page shows all the variables in calculating effectiveness.

Case study in measuring equipment effectiveness

Greenwalt Manufacturing is one of the leading producers of pipe-hangers. They service the electrical and plumbing trades with hangers from 1/2" EMT to 8" Cast Iron Pipe. They are a single shift, 5-day a week operation. All machines are shut down for a 30-minute lunch and 15-minute morning start-up and 15-minute evening shutdown periods.

The company operates punch presses and small press brakes from 12 to 100 tons in semi-automated to fully-automated modes. The firm has its own tool shop and rebuilds presses to its own specifications. Automation is a high priority. A strategy of minimizing setups is used with specialized presses so that changes in set-up only occur once a week (when they need the press for an unusual size pipe hanger, for instance).

This study concerns their new 4-inch riser clamp tooling and set up. Four-inch riser clamps are made from 3/16" by 1" inch mild steel. The component is shaped with a hump in the middle between two ears, so that the body of the clamp will hold 4-inch cast iron pipe securely and the ears will hold up the pipe at each floor level. The clamp is made on a modified press brake in progressive tooling with three stations (form, punch, and cut-off). The machine is fully automatic.

The press brake (Machine number is PB1, Located in the Heavy Hanger Department) is one of the slowest presses in the entire factory, and is rated at 20 strokes per minute. In a typical week there are only three hours of down-time from set up (mostly from material changeovers) and breakdowns totalling (36 minutes/day). Average production for the last 30 working days is 6275/day. This part is a rough type so quality problems are few and far between and are usually related to start-up, problems in the incoming steel finish, or coil ending losses. The company estimates that average reject rates from all sources are 25 pieces per day.

Please note calculations for the overall equipment effectiveness.

Compare these results to TPM standards:

Availability	>90%
Performance Efficiency	95%
Rate of Quality parts	>99%

TPM WORKSHEET FOR EQUIPMENT EFFECTIVENESS DATE:

SIX BIG LOSSES	MACHINE NAME/NUMBER	DEPT/AREA

TOTAL TIME IN DAY OR SHIFT	TIME MACHINE NOT RUNNING	EQ	LOADING TEAM
480	60	=	420

LOADING TIME	LESS	DOWNTIME	EQUALS	OPERATING TIME
420	-	36	=	384

OPERATING TIME	DIVIDED BY	LOADING TIME	EQUALS	AVAILABILITY
384	/	420	=	91.4%

(DESIGN CYCLE TM	×	AMT PRODUCED)	DIVIDED BY	OPERATING TM	EQ	PERF EFFICY
.05	*	.6275	/	384	=	81.7%

OVERALL EQUIPMENT EFFECTIVENESS

AVAILABILITY	*	PERF EFFIC	*	RATE OF QUALITY	*	100	EQ	EFFECTIVENESS
91.4%	*	81.7%	*	99.6%	*	100	=	74.4%

BASIC INFORMATION NEEDED		UNITS
TOTAL TIME IN DAY OR SHIFT	480	MINUTES
TIME MACHINE NOT RUNNING (SCHEDULED)	60	MINUTES/DAY
DOWNTIME FROM SETUP AND BREAKDOWN	36	MINUTES/DAY
DESIGN CCLE TIME	.05	MINUTES
AMOUNT PRODUCED	6275	PER DAY
DEFECT AMOUNT OF PIECES MADE	25	PIECES/DAY

NOTES TO THIS ANALYSIS:

Any implementation of TPM has to face real problems. Following the plans already mentioned will minimize the negative effects. At a recent TPM seminar, supervisors were asked what real problems they would encounter when installing TPM. These problems must be thought about, discussed, and overcome, to have an effective TPM effort.

1. Top management sign-off and support throughout a multi-year TPM process. If your management has a short view they might agree to a multi-year plan and withdraw their support after the first year.

2. Top management gives lip service but does not support TPM with their deeper commitment and their time. How do we get top management to be boosters of the program?

3. Production supervisors might criticize rather than solve problems; they also might complain, and express doubt about the success of the project in front of their subordinates.

4. TPM requires minimal downtime. How do you integrate TPM with customer demands and the sometimes, unreal demands of the forecast and production schedule?

5. The workers might object to the perceived `extra' work by a slow down, increases in absenteeism, letting quality suffer, etc.

6. Where does a small, medium, or large organization get time to do all the training necessary.

7. How do you run TPM in a high turn-over situation? Operators don't stick around long enough to get trained.

8. Where do temps fit into TPM? We use temps to operate machines during busy times.

9. How do you solve the problem of inadequate communication between the production group and the maintenance group? Who will really manage the maintenance part of the operator's job? Why will operations take advice about operator maintenance effort with TPM when they won't now?

10. Is there willingness and is there interest to accept the new roles of the two groups.

11. Operators don't like to clean equipment and neither do maintenance people.

12. Where do our die setters (set-up people) fit into this scheme?

13. Where does quality control and calibration fit in?

TPM and JIT

Total productive maintenance (TPM) is indispensable to sustain just in time operations," says Dr. Tokutaro Suzuki, senior executive vice president of the Japan Institute of Plant Maintenance. In a JIT system, he emphasizes "You have to have trouble free eqiment." Prior to the adoption of TPM, Japanese manufacturers found it necessary to carry extra work in progress (WIP) inventory "so that the entire line didn't have to stop whenever equipment trouble occurred."

Dr. Suzuki defines TPM as "preventive maintenance with total participation." Rather than relying entirely on a staff of maintenance specialists to keep equipment in good running order, TPM pushes the responsibility down to the people operating the equipment. "The concept is that the operator must protect his own equipment," he explains. "Thus the operator must acquire maintenance skills."

However, maintenance experts may still make periodic inspections and handle major repairs. And design engineers also play a big role. They must take maintenance requirements and the cost of equipment failure into consideration when they design the equipment, stresses Dr. Suzuki.

Even in a non JIT environment, TPM can yield impressive benefits, the JIPM Executive says, by reducing losses associated with machine downtime, product defects, and low yields. The goal, simply, is to "maximize the total efficiency" of production equipment. In one Japanese plant, adoption of TPM reduced the number of equipment failures by 97%. In other examples cited by Dr. Suzuki, TPM boosted labor productivity by 42% and reduced losses related to downtime by 69%.

Introducing and Managing Predictive Maintenance (PdM)

The ideal situation in maintenance is to be able to peer inside your machine components and replace them right before they fail. Technology has been improving significantly in this area. Tools are available that can predict corrosion failure on a transformer, thread through, examine, and video tape boiler tubes, or detect a bearing failure weeks before it happens.

Scientific application of proven predictive techniques increases equipment reliability and decreases the costs of unexpected failures. Predictive Maintenance is a maintenance activity geared to indicating where a piece of equipment is on the P-F curve and predicting its useful life.

Although any inspection activity on the PM task list is predictive, PdM is reserved for situations where instruments are used for the readings. In this chapter we will use the common definition of predictive maintenance, which generally refers to technologically-driven inspections. Another way to describe it is to say that PdM is inspections that you have to buy something (an instrument, scanner, etc.) to perform.

Condition-based maintenance is related to predictive maintenance. In condition-based maintenance the equipment is inspected, and based on its condition, further work or inspections are done. For example, in traditional PM a filter is changed monthly. In condition-based maintenance the filter is changed when the difference in pressure between a gauge before the filter and a gauge after the filter reaches a particular value. The differential pressure indicates the condition of the filter.

All the predictive techniques we are going to discuss here should be on a PM task list and controlled by the PM scheduling system. The PdM tasks should be coordinated with other PM activity. Most PdM tasks are handled in string-based PMs.

Maintenance has borrowed tools from other fields such as medicine, chemistry, physics, auto racing, aerospace and others. These advanced techniques include all types of oil analysis, ferrography, chemical analysis, infrared temperature scanning, magna-flux, vibration analysis, motor testing, ultrasonic imaging, ultrasonic thickness gauging, shock pulse meters, and advanced visual inspection. In fact RCM II by John Moubray lists over 50 types of predictive maintenance. Most metropolitan locations have service companies to perform these services, or rental companies to try some techniques in your facility.

Other instruments not discussed here but that should be considered part of your predictive maintenance toolbox are meggers, pyrometers, VOM meters, strain gauges, temperature-sensitive tapes, and chalk. Many techniques and instruments can also be useful in a predictive way because predictive maintenance is an attitude along with a technology! Before you start a predictive maintenance program, consider these questions:

1. What is our objective for a predictive maintenance program?

- ❑ Do we want to reduce downtime, maintenance costs or the stock level in storerooms?
- ❑ What is the most important objective?
- ❑ Will someone else pay for the inspection (like a power utility or an insurance company)?

2. Are we ready for Predictive Maintenance?

- ❑ Do we have piles of data that we already don't have time to look at?
- ❑ If one of the PM mechanics comes to us asking for a machine to be rebuilt, do we have time to rebuild a machine that is not already broken?
- ❑ Would production give us down time on a critical machine on the basis that it might break down?
- ❑ Are we willing to invest significant time and money in training? Do we have the patience to wait out the long learning curve?
- ❑ Are we ready to train at least two people in the techniques (but should consider training almost everyone)?

3. Is (are) the specific technique(s), the right technique(s)?

- ❑ Does the return justify the extra expense?
- ❑ Do you have existing information systems to handle, store, and act on the reports?

- ❑ Is it easy and convenient to integrate the predictive activity and information flow with the rest of the PM system?
- ❑ Is there a less costly technique to get the same information?
- ❑ Will the technique minimize interference to our users?
- ❑ Exactly what critical wear are we trying to locate?

4. Is this the right vendor?

- ❑ Will they train you and your staff?
- ❑ Do they have an existing relationship with your organization?
- ❑ Is the equivalent equipment available elsewhere?
- ❑ If it is a service company, are they accurate, knowledgeable and professional?
- ❑ How do their prices compare with the value received, and to the marketplace?
- ❑ Can the vendor provide rental equipment (to try before you buy); can they provide a turn key service with reports and a hot-line for urgent problems?
- ❑ Does the vendor sell used equipment?

5. Is there any other way to handle this matter instead of by purchase (and capital expenditure)?

- ❑ Can we rent the equipment?
- ❑ Can we use an outside vendor for the service?

In the beginning of the program, the key to all techniques is to have some resources to test your conclusions, and to learn from your results. It must be okay to overreact in the beginning while you are on the steep part of the learning curve. In medicine, these indicators are false positives. If your inspection leads to an unnecessary rebuild, ask yourself, why did this happen? If something breaks down, that you scanned or inspected, ask yourself how did you miss it? These indicators are called false negatives.

Oil analysis

One of the most popular techniques available to measure and predict internal conditions and impending failures is oil analysis. This technique is widely used for engines, large gear-boxes, turbines, compressors, and transformers. There are four basic types of oil analysis, and they are all related to particle size and composition:

Type	Size	Material
1. Atomic Emission (AE) Spectrometry	1-10 microns	all materials
2. Atomic Absorption (AA) Spectrometry	5-35 microns	all materials
3. Ferrography	20-100 microns	ferrous type only
4. Chip detection	40 microns-up	metals only

Figure 41 Oil Analysis

AE, AA

The two spectrographic techniques AE and AA, are commonly used to determine which elements are in the oil. These tests report the presence of all metals and contaminates, based on the fact that different materials give off different characteristic spectra when burned. The results are expressed in PPT (Parts per Thousand), PPM (Parts per Million), or PPB (Parts per Billion). Physical property tests are also run to determine characteristics of the oil such as viscosity.

The lab or oil vendor usually has baseline data for types of equipment that it analyzes frequently. . The procedure is to track trace materials over time and watch the trend. Using the engineering drawings or experience, you can determine where the different elements came from. At a particular level, and given an upward trend, experience will dictate that intervention is required (a re-build or re-manufacture). A typical oil analysis for a stationary engine (such as on a generator set), costs $10 to $25. For big oil buyers, analysis is frequently included at no charge (or low charge) from the oil supplier.

The report is usually in computer-printed form, with a reading of all the materials in the oil and the `normal' readings for those materials. If requested, the lab might call, e-mail, or fax the results so that you can take some action before more damage is done.

For example, if silicon is found in the oil then a breach has occurred between the outside of the machine and the lubricating systems (silicon contamination frequently comes from sand and dirt). Another example would be an increase in bronze particles from 4 PPT to 6 PPT probably indicates increasing normal bearing wear. This wear indicator would be tracked, and could be noted and checked at the regular inspections.

Oil evaluation includes an analysis of the suspended or dissolved non-oil materials including Babbitt, Chromium, Copper, Iron, Lead, Tin, Aluminum,

Cadmium, Molybdenum, Nickel, Silicon, Silver, and Titanium. In addition to these materials the analysis will show contamination from acids, dirt/sand, bacteria, fuel, water, plastics, and even leather.

Other aspects of oil analysis are the characteristics of the oil itself. Questions that need to be answered include: has the oil broken down, what is the viscosity, and are the additives for corrosion protection or cleaning still active?

Consider oil analysis as part of your normal PM cycle. Oil analysis is relatively inexpensive so consider doing it in the following circumstances:

1. After any overload or unusual stress
2. If sabotage is suspected
3. Just before purchasing a used unit
4. After a bulk delivery of lubricant, to determine quality, specification, and if bacteria are present
5. Following a rebuild, to baseline the new equipment, and for quality assurance
6. After severe weather such as flood, hurricane, or sandstorm (assuming the unit got wet or sandy).

Here are two common but simple things to remember with oil analysis. The first is to replace any oil removed for samples. At one large manufacturer, a PM inspector kept forgetting to replace the oil. You can imagine the results! The second common thing to consider is when the oil is being removed. A defense company was doing analysis after oil or filter changes in alternate months. They thought the technique was invalid because the levels varied so wildly.

Other tests carried out on power transformer oil are designed to check the condition of the dielectric and its breakdown properties (a major transformer outage could disrupt your whole facility!).

Oil Analysis Vendors

The first place to begin looking at oil analysis is via your lubricant vendor. If your local distributor is not aware of any programs, contact any of the major oil companies. If you are a very large user of oils, and are shopping for a yearly requirement, you might ask for analysis as part of the service. Some vendors give analysis services to their larger customers at little or no cost.

Labs that are unaffiliated with oil companies exist in most major cities, especially cities that serve as manufacturing or transportation centers. Look for a vendor with hot line service who will call, e-mail or fax you if a warning of an imminent breakdown is found. These firms will prepare a printout of all the attributes of your hydraulic, engine, cutting oils, or power transmission lubricants. The firm should be able to help you set sampling intervals, provide sample bottles, and train your people in proper sampling techniques.

Tip: Send samples taken at the same time on the same unit to several oil analysis labs. See which agrees, which is the fastest, which has the least cost. In all instances, pick a lab that maintains your data on computer so that they can report on trends, and be sure you can get the data on disk or over the Internet for your own analysis.

Wear particle analysis, Ferrography, and Chip detection

These techniques examine the wear particles to see what properties they have. Many of the particles in oil are not caused by wear, and analysis can separates the wear particles out and indicate trends in them. When the trend shows abnormal wear, ferrography (microscopic examination of wear particles) is initiated to examine the cause.

Several factors contribute to the usefulness of these techniques. When surfaces rub against each other they generate very small wear particles. The oil film limits the metal-to-metal contact and the size and quantity of the particles. In the analysis, the particles generated are divided by size into two groups: small <10 microns and large >10 microns. The small particles are normal and benign.

When the oil film is no longer protecting the metal surfaces, and abnormal wear occurs, the large particle count dramatically increases. This increase is the first indication of abnormal wear. When abnormal wear is detected, the particles are examined (ferrography) for metallurgy, type, and shape. These characteristics contribute to the analysis of what is wearing and how much bearing life is left.

The most obvious wear detection technology is a magnetic plug in the sump of an engine. You examine the particles on the plug to see if there are dangerous amounts of chips in the oil. Chip detection is a pass-fail method of large particle analysis. Too many large particles should set off an alarm. Several vendors market different types of detectors. One type allows the oil to flow past a low-voltage matrix of fine wires. A large particle will touch two wires and complete the circuit to set off an alarm.

Old timers had a useful informal test for the quality of wear particles. They would cut open the used oil filter cartridges when they were replaced and rub the captured wear particles between their fingers. Smooth particles were rounded like sand, and likely to be benign. If the particles felt rough, it was time to look closely at the condition of the unit.

Vibration Analysis

Vibration analysis is a widely-used method in plant/machinery maintenance. A study in the city of Houston's waste water treatment department showed $3.50 return on investment for every $1.00 spent on vibration monitoring. The same study showed that a private company might get as much as $5.00 return per dollar spent. The study and the vibration-monitoring project were done by the engineering firm of Turner, Collie and Braden of Houston, Texas.

Vibration is actually three techniques wrapped up into one field

Displacement: The first technique is measurement of displacement. This measurement is the oldest of the three and is still used on slow-turning, large-diameter shafts (less than 600 RPM). Such a shaft produces no readable vibration because of the speed of rotation. Examples might include a bent propeller shaft on a ship.

Velocity: The second development was velocity, or speed, meters. The measurement was in IPS (inches per second) or MMPS (mm per second). It was found that for mid-speed systems such as the 600-3600 RPM found in most factories, velocity was the most useful indicator. Simple set points such as 0.3 IPS could be used to initiate action. At 0.6 IPS the oil film would fail, causing an accelerated failure chain.

Acceleration: The most modern method is to look at the acceleration of the vibration (speed of change of velocity). Acceleration is measured in G's (1G equals the acceleration of 1 gravity or 32 feet per second^2). Acceleration is most used as an indicator at speeds above 3600 RPM.

Modern analysis looks at different frequencies

Vibration analysis measures the changes in amplitude of the vibration by frequency over time. This amplitude by frequency is plotted on an XY axis chart and is called a signature (for a given service load). Changes to the vibration signature of a unit mean that one of the rotating elements has changed characteristics. These elements include all rotating parts such as shafts, bearings, motors, and power transmission components. Anchors, resonating structures, and indirectly-connected equipment are also included.

Many large stationary engines, turbines, and other expensive equipment have vibration transducers built-in. The vibration information is fed to the control computer, which can shut down the unit or set off an alarm from abnormal vibration. The equipment also has computer readable outputs of the data, which allows transfer of the real-time data to the maintenance information system.

Ten steps for a quick set-up of a vibration monitoring program

1. Rent (to start-up) a portable vibration meter for velocity.

2. Train more than one mechanic in the use of the equipment. The mechanics should understand what is being measured and why. Make it a regular task on a task list, or consider training operators in taking the readings. There are several levels to the training (level 1, 2, and 3). It is critical that several peo ple get training so that they can discuss the issues and learn from each other. Occasionally, a trained vibration person will get some experience and then turn professional (become a contractor and take their expertise you paid for with them).

3. Record readings at frequent intervals. Transfer readings to a chart, use the built-in charting function (if you have a sophisticated unit) or use a spread sheet program and let it do the charting for you.

4. Take readings after installation of new equipment to establish a baseline (some manufacturers require this step). Also, take before and after readings when you overhaul a unit. Be sure you take your readings under the same oading conditions.

5. Compare readings and periodically review charts to help predict repair requirements. (You can use 0.3 IPS as a starting point for intervention.)

6. Do repairs when indicated, do not defer them.

7. Note the condition of all rotating elements; keep digging until you can determine what caused the increase in vibration. Make up case studies when the vibration misled you (so it doesn't happen again).

8. Build a file of success stories and tell them to everyone

9. Try different models and different manufacturers' instruments. As appropriate, move into more-sophisticated full-spectrum analysis, and higher levels of training.

10, Buy the equipment that makes sense for your environment. Train widely and trust your conclusions.

Temperature Measurement:

Since the beginning of the industrial age, temperature sensing has been an important issue. Mechanical friction and electrical resistance create heat ($P=I^2R$). Temperature is the single greatest enemy for lubrication oils and for power transmission components. Advanced technologies in detection, imaging, and chemistry, allow us to use temperature as a diagnostic tool.

Today, there is technology to photograph by temperature rather than by reflected light. Hotter parts show up as redder (or darker). Changes in temperature will graphically display problem areas where wear is taking place, or where there is excessive resistance in an electrical circuit. Infrared radiation is unique because it is almost entirely non-interruptive. Most inspections can be safely completed from positions 10 or more feet away and out of danger.

Readings are taken as part of the PM routine and tracked over time. Failure shows up as a change in temperature. Temperature detection can be achieved by infrared scanning (video technology), still film, pyrometers, thermocouples, other transducers, and heat-sensitive tapes and chalks.

On larger stationary engines, air handlers, boilers, turbines, etc. temperature transducers are included for all major bearings. Some packages include shut-down circuits and alarms that are triggered if temperature gets above certain limits. Infrared cameras can be used for a variety of maintenance and production uses. Collect a few applications to justify the purchase.

Specific examples of areas where savings are possible from application of infrared

A hot spot was detected on a transformer. Repair was scheduled off shift when the load was not needed, avoiding a costly and disruptive shutdown.

A percentage of new steam traps, which remove air or condensate from steam lines, will clog or fail in the first year. Non-functioning steam traps can readily be detected and replaced or repaired during inspection scans of the steam distribution system. Breakdowns in insulation, and small pipe/joint leaks can also be detected during these inspections.

Hot bearings were isolated in a production line before deterioration had taken place. Replacement was not necessary. Repairs to relieve the condition were scheduled without downtime.

Cutting tools were scanned continuously in a large machine shop. Infrared indicators detected dullness quicker than could be achieved with other methods.

Possible uses for infrared inspection	Look for
Bearings	overheating
Boilers	wall deterioration
Die casting/injection molding equipment	temperature distribution
Distribution panels	overheating
Dust atmospheres (coal, sawdust)	spontaneous combustion indications
Furnace tubes	heating patterns
Heat exchanger	proper operation
Kilns and furnaces	refractory breakdown
Motors	hot bearings
Paper processing	uneven drying
Polluted waters	sources of dumping in rivers
Power transmission equipment	bad connections
Power factor capacitors	overheating
Presses	mechanical wear
Steam lines	clogs or leaks
Switchgear, breakers	loose or corroded connections
Three phase equipment	unbalanced load
Thermal sealing, welding, induction heating equipment	even heating

Figure 43

Roofs with water under the membrane retain heat after the sun goes down. A scan of a leaking roof will show the extent of the pool of water. Sometimes a small repair will secure the roof and extend its life.

Infrared is an excellent tool for energy conservation. Small leaks, breaches in insulation, and defects in structure, are apparent when a building is scanned. The best time to scan a building is during extremes of temperature (greatest variance between the inside and the outside temperatures).

Furnaces are excellent places to apply infrared because of the cost involved in creating the heat and the cost of keeping the temperature up. Unnecessary heat losses from insulation faults can easily be detected by periodic scans. Instant pictures are available to detect changes to refractory that could be precursors to wall failures.

Ultrasonic Inspection

One of the most exciting technologies is ultrasonic inspection, which is widely used in medicine and has now moved to factory inspection and maintenance. There are two versions of ultrasonics.

In one version, an ultrasonic transducer transmits high-frequency sound waves and picks up the echo. Echoes are caused by changes in the density of the material tested. The echo is timed and the processor or the scanner converts the pulses to useful information such as density changes and distance.

Ultrasonics can determine the thickness of paint, metal, piping, corrosion, in almost any homogenous material. New thickness gauges will show both a digital thickness and a time-based scope trace. The trace will identify corrosion or erosion, with a broken trace showing the full thickness and an irregular back wall. A multiple echo trace shows any internal pits, voids, and occlusions (which cause the multiple echoes).

In the other version the ultrasonic sound wave is converted into the audible spectrum. This technique is called ultrasonic detection. Many flows, leaks, bearing noise, air infiltration, and mechanical systems give off ultrasonic sound waves. These waves are highly directional. Portable detectors worn like stereo headphones translate high- frequency sound into sound we can hear. We can quickly locate the sources of these noises and increase the efficiency of the diagnosis.

In one application, an ultrasonic generator is sealed into a system (such as a large vacuum system). The detector is walked around the system and its piping. You can insert the generator inside a variety of closed systems such as refrigeration piping or a vacuum chamber. Any whistle or noise denotes a leak, loose fitting, or other escape route. Any leak will cause an ultrasonic whistle.

An excellent example of using an ultrasonic generator and detector for inspection is Bandag's tire casing analyzer. This unit is used in truck tire retreading to detect invisible problems in the casings that could result in failures and blow-outs. Ultrasonics detects changes in density, so imperfections in casings are immediately obvious. The transmitter is located inside the casing and 16 ultrasonic pick- ups feed into a monitor. The monitor immediately alerts the operator to flaws in the casing.

Another interesting application of ultrasound is the shock-pulse meter, which reads the film thickness of oil on bearings. Based on the film thickness and the bearing number, the operator can read out the likely life of the bearing from a table.

Idea for action

Ultrasonic grease guns are becoming popular. When the gun is in contact with the fitting, ultrasonic sound waves are transmitted to headphones worn by the operator. The operator pumps grease into the bearing until the bearing quiets down. The system almost eliminates over-greasing and more importantly, gives feedback indication of what is happening inside the bearing. The quieting of the bearing is a motivator to the PM mechanic because he knows he did something "good."

Advanced Visual Techniques

The first applications of advanced visual technology used fiber optics called borescopes. In fiber optics, flexible fibers of highly pure glass are bundled together. Each fiber of glass carries a small part of the picture. The smallest fiber optic instruments have diameters of 0.9mm (0.035"). Some of the instruments can articulate to show the walls of a boiler tube for inspection. The focus on some of the advanced models is 0.3 inches to infinity. The limitation of fiber optics is length. The longest is about 6 feet (but some are longer). The advantages are cost (about 50% or less of equivalent video technology), and the level of technology (they don't require large amounts of training to support).

Another visual technology gaining acceptance is ultra-small video cameras. These tiny cameras are used for inspection of the interiors of large equipment, boiler tubes, and pipelines. The CCD (Charge Coupled Display) devices that produce the images can be attached to color monitors through cables (some models used on pipelines can go 1000s of feet). The technology uses miniature television cameras, smaller then a pencil (about 1/4inch in diameter and 1 inch long), with a built-in light source. Some versions allow small tools to be manipulated at the end of the probe, others can snake around obstacles. These units are used extensively to inspect pipes and boiler tubes.

Internet Cameras are another useful development. These cameras can be put inside equipment and images can be transmitted over the company's Intranet. Some of these cameras have made it to the Internet so that the world can be riveted by the excitement of that actuator (or whatever).

The major disadvantages are cost (high-end cameras are priced at $10,000 to $20,000), and level of support (the cameras require training to adjust and use). As with all technology, costs and support requirements (at least for the low-end systems) are coming rapidly into reach.

The major advantage of the new video technology lies in its flexibility. You can replace the heads or cables and end up with several units for the price of one. A recent design includes a small powered wagon with a lifting gantry, which can scoot down pipes and look all around Els, Tees, and any obstruction. The gantry can lift the camera 48 inches vertically and rotate the camera through 360°.

Video is a good application for contractors. In most major industrial centers there are service companies that will do inspections for a fee. These firms use the latest technology and have highly skilled inspectors. Some of these firms also sell or rent hardware with training. One good method is to try some service companies and settle on one to do inspections, help you choose equipment, and do training..

Other Methods of Predictive Maintenance

Magnetic Particle Techniques (called Eddy Current Testing or Magna-flux)

Eddy Current Testing is borrowed from automobile racing and racing engine rebuilding, and has begun to be used in industry. This technique induces very high currents into a steel part (frequently used in the automotive field on crankshafts and camshafts). While the current is being applied the part is washed by fine, dark-colored magnetic particles (there are both dry and wet systems).

Magnetic fields change around cracks and the particles outline them. The test shows cracks that are too small to be seen by the naked eye, ordinarily, and cracks that end below the surface of the material. The test was originally used when re-building automobile racing engines (to avoid putting a cracked crankshaft back into the engine). The high cost of parts and failure can frequently justify the test. The OEM's who build the cranks and cams also use the test as part of their quality- assurance process.

Maintenance Interfaces: Where Does Maintenance Fit In?

Understanding the interfaces between maintenance and other departments.

The maintenance department interacts with production, purchasing, stores, traffic, and accounting, in addition to production. This section explains the points of view of these departments, their relationships to maintenance, and how to work with them effectively. Never forget that the goal of each department is the increased long-term profit of the organization. Each function approaches this goal with differing assumptions, tools, and world views.

Production interfaces

Naturally, the closest relationship is with the main customer. There are other customers, but the production department is the most important because it is the source of all the company's revenue. In the discussion about world class maintenance, we stated that you cannot go far wrong if you focus on delivering excellent service to the customer.

When looking at the quality of the maintenance-production interaction in a particular organization, keep in mind that personalities and history might be the most important factors. Sometimes the history goes back quite a ways. In one example, the actual slight happened before the arrival of anyone now there!

People have long memories. They keep these memories alive by telling and retelling the great stories. In this way, operations people pass wisdom on to their successors. If the wisdom is about how a maintenance action lost a big customer, you have a problem. In one plastics plant, the dispute occurred before anyone in the current crew was born!

There are formal and informal interactions between operations and maintenance departments.

Maintenance interacts with production on a number of concrete levels.

The Morning meeting: In many plants there is a short morning meeting that identifies the problems and opportunities presented that day. This meeting can be an important communications interface. The meeting may be attended by the production and maintenance managers, or their subordinates. It is essential that this meeting is short and action oriented.

Service requests, work requests, notifications: As explained earlier in this book, the maintenance department finds out about problems through a formal work identification system. This system is the most common interface in plants with formal work management systems. As also mentioned earlier, consider delivering training to any operator who submits work requests.

Page, beep, phoned emergencies: Even in the most formal plants there is a channel of communications for emergencies. Usually these channels are breakdown or line down pages (some plants use 911), or loudspeaker calls for a supervisor or an area mechanic.

Grab by the collar: In plants that are run more informally, a supervisor or maintenance worker might be "grabbed by the collar" and shown a problem area. The maintenance worker would make some notes if it is a large job and simply fix it if it is a smaller job. This procedure is still the standard in small plants with informal systems. It is a serious problem in a larger plant because the maintenance deterioration can get out of control if excessive amounts of resources are switched from important jobs to urgent jobs.

Production improvement teams: Smart organizations (world class) always make a maintenance person one of the members of the production improvement teams because his/her view is valuable for the success of the team. These teams examine an area where some improvement can be made. It could concern operations practices, reengineering, changes to maintenance, or some other factor.

Machine interfaces: Increasingly there are interfaces between shop floor equipment and the work order systems. Up to this point the factory system would set off an alarm and an operator would write the service request. Nowadays the interface is starting to be direct and eliminating the middleman. In sophisticated medical equipment, with OEM or third party service agreements, the machine often is designed to monitor itself. If one of the sensors shows deterioration, the machine's brain places a page to the mechanic or dispatcher responsible for that territory. The service person might show up and replace a part without the customer even knowing that there was a problem unfolding.

Schedule coordination meeting: One of the most important formal interfaces is where the maintenance scheduler and the production scheduler meet and agree on what jobs will be run next week. See coordination (after the planning section) for additional details.

Shutdown meetings: If a shutdown is scheduled, the production department must be involved in many of the meetings. In particular these events are integral parts meetings dealing with scope, plant shutdown, plant start-up, permits, safety, and risk management.

Accounting and Finance departments

The accounting and finance departments are frequently the center of the belief that maintenance is a necessary evil. Certainly any shifts in the way maintenance is delivered will face the accounting hurdle. Accountants will properly ask "does this contemplated change, transformation, or re-evaluation make financial sense?" In the best sense, accountants are professional disbelievers. Accounting should be totally profit and numbers driven.

Some business texts say that the accounting department is the score keeper of the business game. Their role actually goes far beyond score keeping because, in many areas, the accountants are on the field producing profits. In all organizations, the accounting department is the guardian and final arbitrator of what is profitable and what is not.

One continuing issue is that in the eyes of accounting, maintenance is a pure expense, a necessary evil, if you will. In the General Ledger (the bible for accounting for your organization), expenses reduce profit and anything that reduces profit is bad. Unless we create a different view, maintenance can never be anything else to the accountants.

In addition to their gatekeeper role, the accountants categorize every financial transaction that takes place in the organization. This categorization is essential to determine profit or loss. Maintenance departments are frequently on the cutting edge of these decisions. For example many accounting discussions deal with the decision to capitalize or expense a major repair. In one instance, the repair is an investment in the organization and becomes an increase in the value of an asset. In another example, the same repair is an expense, a nail in the coffin of profit.

Another complication in the relationship is the difficulty of coming up with hard numbers for returns from maintenance investments. We could have a clear improvement (such as the efficiency of a boiler in a plant with several users of fuel) that is almost impossible to prove with hard numbers without a lot of extra work and data collection.

The accounting department uses double entry bookkeeping techniques, which means that every transaction has two sides. A spare parts purchase will increase A/P (accounts payable - what is owed to vendors) when the invoice is posted and will also increase the Expense account. When a machine is purchased, the invoice is booked to A/P and to an Asset account.

All maintenance work must be accounted for within the General Ledger (GL). The accounts mentioned above (expense, asset, A/P) are accounts in the general ledger. The GL is the score card of the organization that reports the condition of the finances, profitability, assets, and liabilities. At the end of a financial year, a snapshot of the General Ledger is compared to the snapshot made last year. The difference is profit or loss.

A repair to a machine is fundamentally different from a rebuild/improvement to the same machine. A repair is an expense that reduces profit and an upgrade or rebuild increases the assets and the profit. These different types of transactions must be differentiated and reported.

Some categories that accounting might be tracking:

- ❑ maintenance parts and spares such as pumps and repair parts
- ❑ maintenance supplies such as rags and absorbent pigs
- ❑ open stock items charged to specific jobs
- ❑ maintenance engineering charges both inside and outside
- ❑ contract labor
- ❑ specific projects, particularly if they have budgets
- ❑ maintenance labor, fringe benefits, overtime
- ❑ labor and materials for grounds maintenance
- ❑ janitorial services
- ❑ maintenance support expenses for supervisors, staff, and managers
- ❑ costs to fabricate tools, jigs, fixtures, and whole machines
- ❑ non-recurrent labor on large repair jobs
- ❑ structural repairs to buildings
- ❑ contractor charges on large repair jobs costing over $50,000

The accounting department also verifies that the transaction doesn't exceed any of the predetermined limits (if limits are exceeded then special authorization might be needed. These permits are also called grants of authority). Example amounts that a maintenance department can spend at a large manufacturer without special authorization.

Type of charge	Limit before higher authorization is needed
Office furniture	$500
Computers, instruments	$1000 (or all computers or software must be approved by IT)
Small tool purchase	$500
Capital spares program purchase	$5000
Any MWO is eligible for capitalization	$5000
$ limit on any individual MWO	$40,000

Costs need to be charged to the department that incurred them. This work of tracking and charging all costs is the function of the cost accounting group within accounting. It is difficult to divide maintenance up into maintenance costs per product line because some costs span several products.

Cost Accounting is one of the most complex functions of the accounting profession because it determines the factory cost of each product line. To do cost accounting correctly, overheads (such as energy, phone, and support departments such as maintenance, purchasing, receiving, etc.) must be allocated to the most logical product. For example, a plant that manufactures wire harnesses uses injection molding equipment. The molders use large amounts of energy to melt the plastic. In fact, the molding department might use 1/2 or 2/3 of all of the electricity used in the whole plant. Cost accounting might calculate this ratio and add the additional overhead to the cost of each plug molded. This procedure assigns costs to products more fairly.

Many fights are the result of illogical or incorrect assignments of overhead. A low-tech division of an aerospace company was stuck with the burden rate of the entire company (which made it competitive in aerospace but non- competitive in the low-tech area.) They could not convince the accounting management that the sophisticated engineering, computerization, and other elements of aerospace made no sense in the low-tech arena.

Charge-out rate

Accounting also maintains charge-out rates for all job classifications. The charge out rate includes hourly wages, fringes (retirement costs, insurance, FICA, vacations, etc.), and overhead (supervision, clerical costs). This number is important because it is the comparison number for outside contractors who are asked to bid jobs against inside people.

One essential duty of accounting is to calculate and supply us with the true Cost of Labor

The true cost of labor includes

1. Direct wages
2. Costs of benefits including health insurance, FICA (employer's contri bution), pensions, workmen's compensation, life/disability insurance, and any paid perks.
3. Indirect costs include indirect salaries (of all support people), materials not charged on work orders, supplies, costs of the shop and tools, allocation of costs of corporate support, costs of money, and hidden, or other indirect costs.
4. A factor for time paid for but not used on chargeable repairs such as vacation, leave, training, meetings, shop cleanup, and idle time.

For example the cost of 1+2+3 from the above list might be:

Direct hourly wages	$25.00
Cash Benefits @ 30% of wage	$ 7.50
Indirect costs	$10.00
Total cost per hour	$42.50

This number would be useable if every one of the 2080 hours each year was worked on chargeable jobs. However, that ideal condition is not normally achieved, so the total cost per hour must be increased to cover time that is paid for but that doesn't appear on any work order.

2080 hours straight time (52 weeks * 40 hr./week) available per year.

Calculations show:

Annual straight time Hours: 2080

Deductions

Vacation	<160> (4 weeks)
Holiday	<64> (8 paid holidays)
Paid sick and personal leave	<24> (3 sick and personal days)

Meetings (safety, scheduling)	<52> (1 hour a week)
Training	<40> (1 week a year)
Other: jury, National Guard, union	<24> (3 days per year for all other categories)
Indirect time (cleaning shop, etc)	<104> (2 hours a week)
No assignment –idle	<52> (1 hour a week)
Total	<520>
Annual time accounted for	1560 hours 75%= (1560/2080)

The factor is 75% (only 75% of the time we pay someone for is time that can be assigned to chargeable jobs)

If we start with a $25 direct wage:

Actual cost of time = $42.50/ (0.75) = $56.67 (We end up with $57/hour rounded charge rate)

Another way to calculate the real cost is to use a 2.28 factor multiplied by direct wages to arrive at the charge rate.

Charge rate	=	Direct wage rate	*	Charge ratio
$57	=	$25	*	2.28

All work orders in that shop with the above assumptions should show $57 as the charge rate.

Note that a maintenance contractor with a $25 wage rate might establish $60 as the floor rate and prefer to charge $75 per hour for a full profit. On one hand, the contractor needs to cover some business costs (like bad debts), and general and administrative costs. On the other hand, generally a contractor will have a higher work ratio than 75% (closer to 90%), and fewer benefits.

The charge rate should be recalculated annually after costs such as labor, insurance, workmen's compensation insurance, and tax rates are known. Normally overtime is charged at the same charge rate because the overhead pool does not increase and the benefit costs (such as health insurance costs, life insurance, etc) don't increase linearly with increased hours. Also, the raw amount of deducted hours doesn't vary much with increased overtime.

Accounting and the Maintenance Parts Inventory

The hardest thing to remember is the impact of the fact that according to the rules of accounting, the maintenance stores do not exist as an asset of the organization. All financial transaction in the organization follow the rules of double entry book keeping, so let's trace two different store room transactions. This discussion is not true for some large organizations, who often treat the inventory as a separate business unit and charge the part when it is taken out of the stock room.

All companies- Accounting steps for raw material purchases:

1. Raw materials come in the door, invoices come in the mail.

2. Invoices are booked to Accounts Payable (a liability account), raw materials are booked to Raw Materials (an Asset account). The two bookings balance each other.

3. An invoice is paid and the amount is booked to the cash account (reducing an asset) while Accounts Payable is reduced by the same amount. Once again the two bookings balance each other.

4. The result is that the asset (cash) is traded for the asset (raw materials).

5. There is no effect on profit until the raw material is used and the product is completed and shipped to the customer or the warehouse. Then the asset is charged to cost of goods sold on one side, and to accounts receivable (or finished goods inventory) on the other.

Most companies- Maintenance parts purchase:

1. Maintenance parts come in the door, invoices come in the mail.

2. An invoice is booked to Accounts Payable (a liability account), and the related Maintenance Parts are booked to Expenses, specifically maintenance parts and repair (an expense account). Since the account named maintenance parts is an expense item, the company profit is reduced immediately.

3. An invoice is paid and the amount is booked to the cash account (reducing an asset) while Accounts Payable (liability account) is reduced by the same amount.

4. The result is that the asset cash is consumed by an expense (maintenance parts and repair).

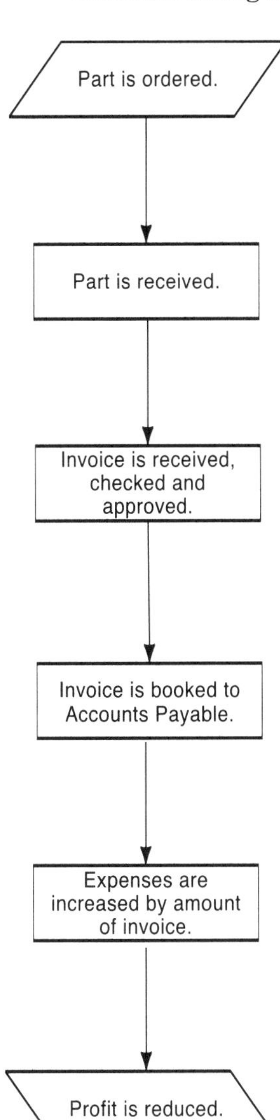

Figure 45 How parts are charged against the books.

5. The item might be sitting on the shelf in the store room but there is no asset on the books.

When the organization needs more cash or additional profit, one place to look at is the maintenance parts in the store room. These parts are an asset that does not appear in the books. Metaphorically, if you squeeze the store room, profit and cash drip out! This conundrum explains the zeal with which the maintenance stock is attacked.

Effective arguments to use with accounting:

If you want to interest accounting, you have to talk about topics in which they are interested. Don't expect these discussions to go well at first. By its nature (and probably for good reason) accounting will distrust your numbers until you start demonstrating competence.

Savings acceleration: Maintenance savings flow directly to the bottom line. The profit of a typical organization is $0.05 to $0.10 for every dollar of revenue. A dollar of maintenance cost avoidance or cost reduction is worth $10 to $20 of additional revenue.

Cost reduction: The maintenance budget was $7,000,000 last year and will be $6,500,000 this year with these changes. . .

Cost avoidance: At the present rate of deterioration our costs will go up $500,000 next year unless we take these steps.

Cost control: The maintenance effort cash needs have varied by 50% month to month for the last 3 years. With these changes we will be able to predict the cash needs to plus or minus 10%...

Increased production through increasing the availability of assets: We achieved 7950 hours of production last year on the big widget line. With the proposed modifications and improvements we will run 8220 hours next year.

Insurance policy as a model for capital spares: The failure of any of the 30 parts on this list could shut down production for at least 4 weeks and perhaps as much as 12 weeks. We lose $1,000,000 of direct revenue each week we are out of production. We currently stock only five of these parts. For an investment of $375,000 we could stock the rest of the parts and insure the uptime against catastrophe. Like an insurance policy, we buy it not to use it. In other words, as with fire insurance we would buy the part and hope not to use it! Also, like fire insurance, we would conduct risk management exercises (in maintenance they are called PM or PdM) to reduce the probability that we would ever need the part!

Maintenance has impacts throughout the organization. Its impact affects the entire cost structure of your product. When talking to the accounting department, specific causes and effects must be mapped. Many costs saved by good maintenance practices are outside the domain of the maintenance budget. Look for costs outside the normal maintenance area for persuasive arguments.

Cost areas to discuss:

- ✓ Maintenance Parts, Labor,
- ✓ Contract labor, service contracts
- ✓ Energy cost including electricity, gas, oil, coal
- ✓ Water costs
- ✓ Costs of shorter equipment life
- ✓ Costs of downtime
- ✓ Costs of scrap
- ✓ Costs of poor quality (variation in the process)
- ✓ Costs of accidents- medical, lost time, legal
- ✓ Costs of environment fines
- ✓ Costs of lowered productivity

Purchasing

The purchasing department's mission is to buy materials necessary for the business that meet the specifications at the lowest cost. In addition, purchasing usually is involved in contractor negotiation, new machine procurement, and sometimes is in control of the storeroom.

Maintenance department demands are a headache for most purchasing departments. Maintenance demands make it very difficult to implement continuous improvements in purchasing. The needs of maintenance are at odds with some of the methods usually used for improvements in purchasing. The total cost to buy an item is called the COA (Cost Of Acquisition).

Continuous Improvement Techniques	**Maintenance reality**
Reduce the number of vendors	100's of vendors
Become a bigger fish for better service	Small buys
Planned purchases	Unplanned events
Consolidate buys, blanket orders	Uneven requirements
Take time to do it right 1st time	No time, rush, ASAP
Time for low cost shipment	Air freight, courier required
Maintain quiet, efficient atmosphere	Crazy, emergency room atmosphere

Figure 43

The purchasing department is one of the major interfaces to the outside world. It also is charged with the mission to protect the organization from certain types of lawsuits and fraud. In the US, all organizations are regulated by a series of commerce laws called the UCC (Uniform Commercial Code), which regulate commerce. Additional coordination and logistical responsibilities are added when the firm is a JIT (Just-in-time) manufacturer.

Every dollar saved by purchasing flows to profit. In some organizations, the major portions of funds are spent by purchasing for raw materials and supplies. This direct relationship with profit sensitizes purchasing agents to saving money. Losing control, even of dimes and quarters, can affect profit by thousands of dollars at the end of the year. Unlike maintenance, which labors without oversight and whose actions are evaluated indirectly, the actions (or inactions) of purchasing are sometimes painfully visible to accounting and to organization management.

The main techniques used to get the best price, specification, and delivery, are shopping and contract negotiations. Because of the requirements of the organization's auditors and legal counsel, the preference is to shop around and get bids or quotes from several sources. In public agencies, this process is required by law for contracts over a certain dollar amount. In most factories the same practice is strongly recommended by customs and regulations. Fair (and well documented) shopping is the best protection the Purchasing Agent has against charges of fraud, collusion with vendors, and other malfeasance.

Because of shopping, the purchasing process takes time. Preparation, workload, complexity of the buy, experience, and quality of history files, impact the time it takes to issue a Purchase Order (PO). This time and effort is summed up in the cost of acquisition. There are several levels to the cost of acquisition but the more complicated the buying procedures, the higher the cost of acquisition. Of course, technology is lowering the costs of acquisition.

To a maintenance professional, doggedly shopping for each part looks like a waste of time. Shopping is good purchasing practice. Good mechanics will not necessarily fix the first thing they see but will wait until they understand why the failure occurred. That approach is good maintenance practice; it takes a little longer but in the long run makes sense.

The other problem that seems to occur with regularity is that the maintenance planner or supervisor locates the part that is needed. To locate the part, they work with a vendor (who might put significant time into the research). The requisition, with details of all the research and the vendor recommendation are sent to purchasing.

Purchasing then ignores the recommendation and buys the part from another vendor. This sequence causes an unfortunate communications problem. Meetings with the purchasing agent and other managers should be held to discuss and determine policy in these situations. Sometimes the value added is worth the higher price and sometimes it amounts to a sweetheart deal for the vendor.

In a purchasing department for a metal producer, over 13,000 purchase orders are issued in a typical year. In a recent downsizing, their staff was halved. The reaction of the survivors, which will be hailed by maintenance departments, was to set-up 30 prime maintenance vendors and several thousand commonly-used items. They called these items `E` items. The vendors had pre-negotiated blanket contracts for `E` items. The maintenance planning department was given the authority and responsibility for buying E items from these vendors, and they had a limit of $500 per line item.

The move toward JIT adds to the complexity of the purchasing mission and increases the stakes. Traditionally, buffer stocks insulated the factory (and the purchasing department) from costly stock-outs of critical raw materials. JIT techniques eliminated the buffers. Now, when delivery problems arise, purchasing is on the front lines to contribute to solving any problems.

When talking to maintenance managers, one constantly hears stories about the purchasing department buying junk to save a few dollars. Purchasing uses specifications prepared by engineering, production, and maintenance to determine what it buys. Specifications are detailed descriptions of the qualities, performance, and functions of the items purchased. Usually these mistakes are traceable to the specifications being incomplete or incorrect. Purchasing is totally dependent on the completeness and accuracy of the specifications. They cannot, independently, judge the qualities of the parts or supplies purchased.

Over the years, organizations make mistakes and learn of things to avoid. The Purchasing department (in conjunction with legal counsel) keeps track of this learning in its `Terms of purchase' (which are usually printed on the back of the P.O.). Some typical terms and what they are designed to protect against are :

Acceptance: The P.O. sent to the vendors is just an offer to purchase. The vendors have to acknowledge the P.O. or act as though they accept, by shipping the order, for example. The P.O. is not tied to a quote or bid by the vendor but is an offer that they can accept or reject. In legal terms, acceptance actually forms an enforceable contract. It is important to know when and if the PO was accepted.

Changes: Must be agreed to in writing to be binding. Verbal changes are the subject of a large proportion of lawsuits with vendors. This provision protects both organizations.

Force Majeure: Either party can be excused from performance by acts of God, fires, labor disputes, acts of government, and a whole list of possible problems. After a catastrophe, the last thing you need is a flurry of lawsuits for failure to deliver parts.

Design Responsibility: The seller is totally responsible for the design of the product. The P.O. is a request to meet specifications, not a design of a product. If the design is defective, the blame is placed on the vendor, which protects your organization from certain liabilities.

Warranties: Any products bought by P.O. are warranted by the vendor as fit for the intended purpose and should be free from defects. The seller must fix and/or replace defective items, or refund all money paid, including shipping charges. Products should do what they are supposed to do, and if they do not, you can return them.

Intellectual property rights: Sellers own the rights to the product they are selling. The seller is not infringing on anyone else's patents or licenses. This aspect protects you against the suit in which the vendor is in dispute with a third party about patent ownership of the item purchased. Some third parties might sue you if the purchase is large enough.

Confidentiality: Any information told to the seller, or anything developed by the seller for you, is confidential. This confidentially lasts for x number of years. You might have to tell a vendor a trade secret, or show them a proprietary process, so that they can help you. This clause alerts them to keep the secret. Where critical secrets are revealed, stronger non-disclosure agreements are usually required.

Termination: For stock items you can cancel the order and pay only for items shipped. For a custom item you will pay only for the work accomplished. Certain overhead costs and profit might be paid after negotiation. Termination before completion of an order is always a sticky affair. This clause limits your liability.

Indemnity: Except for negligence on your part the seller indemnifies (protects) you from any damage (lawsuits) brought as a result of the purchase. If a batch of chemicals you purchase explodes while being mixed at the vendor's facility, you are protected by the vendor from being held liable.

Assignments and sub-contracts: A P.O. cannot be assigned, or the work sub-contracted, without written consent. You want to know who is actually doing the work.

Delivery: The promised date is important (time is of the essence). If a vendor fails to deliver in the specified time you can cancel the order without any damages or buy the product from another vendor and charge the first vendor the difference. For some items, delivery time is critical. This provision alerts the vendor that you can get damages from them if they miss deliveries.

Price warranty: The seller warranties that the price charged is the best price charged to any customer for this volume. The price shown on the P.O. is complete and there are no hidden charges. This provision prevents the vendor from charging you a higher price than anyone else buying that quantity. Hidden charges have to be un-hidden to be acceptable.

Default: If the vendor breaks any of the rules you can break the contract without any additional charges. This legal device is designed to allow you the greatest flexibility to cancel if the vendor breaks any rule.

Ideas to improve the purchasing maintenance relationship

1. After an emergency is over, invite the purchasing agent or buyer into the shop to show them what was bought. Frequently a $35-solenoid is very impressive when actuating a $700,000 machine that generates $48,000 a day.

2. Give time when possible. Many emergencies are due to a lack of planning on your part. Work on getting the maintenance act together to reduce the number of unscheduled events.

3. When you or your staff do research into a vendor for a part, pass along your work but do not expect that it will be followed exactly. If the vendor puts significant time into the project, bring the purchasing department into the loop early in the process. In some circumstances it is fair to pay the vendor for the work and bid the requirement with an even playing field.

4. Look closely at specifications on items that have been giving you trouble. It is possible that your problem is in the specs and not in the purchasing department.

5. Humanize the relationship by looking at some of the people in purchasing. Find out what makes them tick and what pressures they face.

6. In larger departments, try to negotiate a specialist to handle the bulk of the maintenance buys. The best solution may be to move the specialist to the maintenance department.

Saving money in maintenance purchases

To analyze inventory: sort each part number by dollar volume (unit price times yearly usage). In one study it was determined that the top 7% of the line items represented about 75% of the yearly purchasing volume. These high- cost and fast-moving items are called the `A` items.

1. Identify `A` items.

2. Apply rigorous purchasing and negotiating techniques to `A` items to lower costs. This exercise is an excellent application for the skills of the purchasing department.

3. Review specifications to get better parts at the same costs or equivalent parts at lower costs. Review of the specifications might reveal a more expensive part that lasts longer (see #6 below). In such a situation, some level of engineering can pay off in big returns. Areas to look at can include lubricants, bearings, wear parts, fittings, belts, fasteners, and wiring/electrical.

4. Apply sophisticated standards to setting reorder point and economic order quantity. Factory production inventory control experts have long studied inventory strategies. `A` level items respond many to these techniques.

5. Consider creative new vendors, purchasing modes, and approaches such as going directly to manufacturers instead of purchasing fasteners from industrial vendors. Another significant way to save is to buy bulk (purchase truckloads of a high-volume oil, of course you'll have to installing tanks but it might be well worth it).

6. Although not a savings for purchasing, these `A` items also represent most of your labor cost. Any successful attempt to reduce the need for these parts will also reduce the need for your labor. In other words, the `A` items provide the best opportunities for maintenance improvement

Stockroom, stores, and the maintenance inventory

How much do inventory items really cost? Sometimes not having the part is a better business decision. Where there is a good vendor and downtime is not a severe constraint, paying a little more to not stock the item might be a good practice. Numbers to calculate the real cost of inventory should be available from accounting or from the warehouse itself.

True Cost of Inventory (called Cost of Ownership or COO)

Consider the distinction between the price paid for an inventory item and the selling price of that item of inventory. A retail parts store would add certain real costs plus profit to determine minimum selling cost. We don't have to add profit but we do have to add certain real costs, the costs of having and using inventory. These costs have to be spread over the parts used and added to the part purchase price to yield the part charge out cost.

These costs are referred to as the Costs of Ownership (you'll see them as COO in the financial reports). Inventory ownership costs, which are unique for each operation and can even be unique for each location within an organization, include the items discussed below. Mark Goldstein uses 2.5x prime rate as a COO rule to thumb.

1. Cost of Money: Your organization could invest the money now tied up in inventory at market rates and get a secure yield from 2% to 10% per year for the value of the inventory in today's market. The charge is linear so large inventories are not an advantage (as they are in some other line items).

2. Expenses of Warehousing: These costs include depreciation on building space and shelves, an allocation of costs of utilities, building maintenance and security, life cycle costs on material handling equipment, forms and paper, office supplies and machinery. Cost is usually figured at 2.5% to 6.5% of inventory value. Larger inventories would have lower relative warehouse expenses.

3. Taxes and Insurance: Some localities tax the assets of the organization, and most have real estate taxes. Casualty insurance must also be included. Costs vary from 0.5% to 3%. The charge is linear so large inventories are not an advantage (as they are in some other line items).

4. People Costs: Full time stock clerks, allocation of other clerks, pick-up/delivery people, supervisors, and purchasing agents. Overall parts volume has a great influence and the cost is from 10% to 25% of the annual volume. these costs are not directly related to the amount of the inventory but to the volume of transactions. Large numbers of small transactions are almost as expensive as the same number of large transactions. Again, larger warehouses would have lower relative people expenses. Also, highly automated warehouses are less expensive than manual ones for the same number of transactions.

5. Deterioration, shrinkage, obsolescence, and cost of returns: Parts sometimes become unusable, obsolete, disappear, must be returned for

warranty, or incur a re-stocking fee. Costs vary from 4% to 15% (higher if you use a lot of common items where shrinkage is a significant problem). Usually larger inventories would have lower relative deterioration, shrinkage and obsolescence expenses.

For a $600,000 inventory turning twice a year with a 26% (low) COO ratio:

Best case: $600,000 * 26% = $156,000 (COO) 1 + $156,000/$1,200,000 = 1.13

For a $600,000 inventory turning twice a year with a 54% (high) COO ratio:

Worst case: $600,000 * 54% = $324,000 (COO) 1 + $324,000/$1,200,000 = 1.27

On this basis, a $500 machine part would be charged out between $565 and $635

Why have a maintenance storeroom?

The primary reason for a maintenance inventory is to help increase uptime and raise production volume. The secondary reason is improved maintenance worker productivity. Between 40% and 60% of all maintenance dollars flow through the store room. Having the right items in the right quantities is the challenge. The storeroom also can help improve maintenance productivity (and reduce downtime) by being able to find any part in stock within a few minutes.

Plenty of heat to go around

The store room takes heat if there is too much stock (from management), too little stock (from maintenance, from production), if a part is not in stock that is on the computer (from maintenance, auditors, and production), if it takes too long to find parts (from the mechanics), or anything else that goes wrong. Next to purchasing, the store room is the most complained about department servicing maintenance.

Types of inventories

The retail model of store keeping: The most common type of inventory is a resale inventory. In this model, money is made by selling the inventory and replacing it at a lower (wholesale) cost. The goal is to have exactly the amount of inventory that your customers want. This model says that every item that does not move in a given period of time should be removed from the shelf and eliminated. If something doesn't move then it should be replaced by something that will.

Retail inventory models are taught in business school and are well understood by accounting, purchasing, finance, and marketing. The problem that this arrangement presents to maintenance is that large numbers of people have expertise in an inventory type that is very different from most of the inventory held by the maintenance storeroom. So people dealing with maintenance inventory have experience and training with a very different kind of inventory problem. The result, for maintenance, is that it is very difficult to get through to them with new ideas and constructs because they already know about inventory.

The ruling ratio is turns of inventory per year. That ratio looks at the total amount of purchases that year, divided by the inventory left at the end of the year. Each area of the store is also looked at. In a retail store, every square foot of shelf space is evaluated for its productivity. The metric is sales per square foot (I have not heard of anyone applying this to maintenance. . .yet!).

Companies like Walmart have turned this type of inventory management into a science. In the Walmart system, when a family checks out at the cash register all their items are recorded against the inventory. So far, everyone in the business does that. But Walmart goes farther. The demand is also transmitted to the manufacturers, so that, when a case of soup is sold, the demand is sent directly to the soup vendor. The vendor (not Walmart) owns and manages each item in the Walmart store's inventory.

The Walmart store holds the product on consignment and pays when it is sold. The soup vendor accumulates the sales for each store and delivers palletized loads of exactly the right amount of each different product for each store. The pallets are delivered to the Walmart warehouse where they are unloaded from the incoming truck directly to the outgoing truck. Products are said to never touch the ground (this procedure is called cross docking).

The point is that Walmart has a lower cost of ownership than anyone else, which gives them the advantage of having a lower cost of goods. Walmart translates that factor into lower sales prices. The low sales prices encourage high volume so the cost of the high technology is amortized over hundreds of billions of dollars of annual sales.

Production inventories: Another type of inventory is the production inventory. This inventory consists of raw materials and subcomponents that are consumed by the production process. The production flow is the most important issue. In older plants there would be weeks or months of parts, components, and raw materials waiting for production. The costs were enormous.

Also, quality problems caused complete havoc on the line because if there was a defective part there might be thousands of them to use up or scrap before the problem was identified and resolved.

Before Harley Davidson changed their manufacturing process they maintained a raw material inventory that was in-house an average of 17 weeks before it was used. After they changed their process the raw materials sat around a week or two at the most. Many items came in and left within 3 days. The measure in this kind of inventory is in days (or hours) of stock and work in process. The goal is zero work in process. In fact, many plants have managed to run with only a day's production buffered.

The advances in this inventory area have caused problems in maintenance. JIT (Just In Time) works well in production where you can predict the needs of the process. It creates a problem because there is no buffer when a machine breaks down. JIT usually requires an increase in the maintenance stock level, as well as improvements in reliability.

Maintenance parts stores: The type of inventory we deal with in our stock rooms is in support of productive assets. It is unlike production or sales inventory because the goal is to be able to insure equipment availability. We want to keep the minimum stock level that will support the equipment. If we had a maintenance parts distributor next door with good prices and all our parts, we would not need an inventory at all.

The value of the loss of production from not having a part is usually orders of magnitude greater then the cost of the part even with the costs of ownership.

Of course, we will do everything in our power to not have unnecessary parts on our shelf. We want to set up valued vendor programs to lower stock. Some vendors have instituted consignment stock (somewhat like Walmart with a few important differences). In the maintenance model we pay when the parts are replaced (not when they are used). This difference is important because any shrinkage is our problem. Consignment stock in maintenance is designed to reduce cost of ownership while increasing convenience and lowering inventory.

We also want to keep the money invested as low as possible. The goal is production. The measure should be lowest downtime due to parts shortages.

Functions of the store room

Administration:

- Provide information to the planner or supervisor (are items on the shelf?)
- Provide usage information to purchasing (for economical order quantities)
- Calculations associated with stock keeping (min, max, EOQ, safety stock)
- Maintaining the parts' numbering system
- Creating and maintaining parts catalog
- Preparation of reports for management
- Tracking parts to and from rebuilders
- Account for all parts received, used, and left on hand

Receiving:

- Unload trucks
- Count parts
- Inspect parts for compliance with specifications
- Facilitate quality efforts
- Check for correct part numbers
- Verify P.O. exists and details match packing slip
- Provide proof of receipt to accounting to pay vendors.

Storage:

- Put away parts
- Store without damage
- Rotate stock
- Provide security against theft and vandalism
- Provide proper storage environment
- Count parts (maintain physical inventory)

Maintenance support:

- Reserve (put aside) parts for jobs
- Manage rebuildables, both incoming and outgoing
- Build Kits with parts, supplies, and possibly tools for common jobs
- Pull parts for jobs off shelves and prepare them for transportation (or pick-up)
- Locate parts
- Help with research into infrequently-used parts
- Identify unknown, broken parts

Traditional Problem areas to address:

- Hoarding of parts by craftspeople
- Field re-engineering (with no documentation)
- Vendor substitution of a different part with the same part number
- Inadequate storage
- Slow parts window service
- Inflexible and hard to use computer systems for inquiries and research
- Inadequate tracking and staging of parts coming from and going to re-builders
- No physical inventory taken, all kinds of stuff in the inventory
- No one wants to write-off bad or obsolete stock and take the hit
- Bad relationship between stores and the tradespeople
- No ability to add to or remove from the stock list

Three categories of part uses

1. Usage-driven parts

These are the simplest type of part to understand because these are the parts that wear-out, (belts, bearings, wear parts). You use the part because it wears out. If your production goes up, usage of these parts usually increases somewhat in step. If production goes down, usage eventually goes down. If you get a slew of novice operators, usage of these parts increases until everyone has been trained. This type of part is usually well understood by management outside maintenance. Usage-driven parts respond well to retail inventory analysis such as evaluating by TURNS or EOQ. In fact, this kind of part usage is well understood because it is most like the retail inventory.

2. Paperwork-driven parts

Parts used for construction, PM and PCR parts. Usage of the part is driven by a number on a drawing or a study of paperwork. The problem is that any change in the paperwork immediately changes the part needs. This type of part can be disruptive to a stock keeping system because the usage can change wildly as the paperwork is modified. For example, look at the impact of a large construction job on EOQ (annual usage will be artificially inflated). Many systems exclude this kind of usage from calculations.

3. Fear-driven parts (also called Insurance Policy Spares or Capital Spares)

Fear-driven parts have long lead-times and are hard to duplicate for machines or processes with high downtime costs. These parts support mission-critical machines. They may come from overseas. In a power utility, these parts are called Capital Spares, and they are handled differently on the books. When a new plant is built, spare parts are purchased and depreciated along with the machines on which they are used.

Another name is insurance policy parts. These parts are regarded as an insurance policy against downtime. Like insurance policies, they have premiums (the cost of the part and carrying cost of the store room - COO), benefits (avoided downtime). Fear-driven parts do not respond to the retail model and tend to confuse conventional inventory analysis.

Remember, the fear parts are on the shelf (just like your fire insurance policy) to **not** use. In fact, using your insurance parts usually indicates of a breakdown in your PM system. Just as with insurance, you take all possible measures so that the part (policy) is not needed. But also like insurance, you are grateful when you have a major breakdown and have the right parts (you bought the insurance and now it pays off) to get the equipment back on-line.

Primer on Maintenance Storekeeping

Specialized vocabulary of the storeroom

Annual usage	How many do you use in a year	Obsolete part	Part you don't need any more or that have been superseded
Bin Maximum (or just plain-Maximum	What is the greatest number you want on the shelf	OEM	Original Equipment Manufacturer
Carrying cost	COO already defined	Open stock	Items you can walk up and take without paperwork, such as bolts
COA(cost of acquistion)	All the costs to buy and receive something	Opportunity cost	The return on investment that the company could realize with the money tied up in the parts room
Consumables	Items that are consumed in the production process such as printing ink, cutting blades, etc.	Order Cost	Cost to process 1 order
Critical spare AKA Capital spare and AKA Insurance Policy Part	Critical, high cost, difficult to get part	Restocking cost	Cost to return an item to the vendor
Cycle counting	Staggered counting of some parts every week.	ROP (reorder point)	What inventory level to reorder also called minimum
EOQ	Based on balancing the COA with the COO what is the best number of items to buy at a time.	Safety stock	The difference between the average use of a part and the maximum use of the part for the lead time involved
E-MRO	Usually Internet-based parts buying and research	Stock item	Items you keep on the shelf
Expedite point	Point where you call the vendor to find out where the shipment is	Stock Out	When an item should be in stock (it is a stock item) and isn't
Lead Time	How long it takes to get a part. Should include the internal lead time (time to place an order).	Supplies	Rags, oil dry, gloves – items used but not charged on work orders
Maximum usage	Maximum probable usage of a part per week	Turns	Purchases divided by inventory value. Measure of how active the inventory is.
Minimum	Same as reorder point	Unit cost	How much does one part cost
MRO	Maintenance, Repair, and Overhaul		

Figure 48

How to conduct a physical inventory: There are two methods of conducting a physical inventory. The old method is annual physical inventory where the department shuts down for one or more days and physically counts everything. Newer methods use cycle-counting strategies, which means that 1/12 of the inventory is counted every month. In a year, all the parts are physically counted. The advantage of cycle counting is that no shutdown is necessary. Cycle counting can also be accomplished with existing staff during slack periods. You may also have a great deal more flexibility with cycle counting if you start having stock-outs. In addition to cycle counting the counts are updated whenever there is a stock-outage.

There are more sophisticated methods of cycle counting in which the parts that move quickly or are more critical, are counted more frequently and the slower and less critical parts are counted less frequently. With these methods, fast-moving items might be cycle counted quarterly and slow-moving items counted every 24 or 36 months.

Formulas

Many formulas are used to determine stock levels and reorders. All these formulas are based on procedures used in retail inventory sectors so they will not easily apply to insurance policy items. If a critical spare is needed once in 6 years, and it takes two years to get, conventional retail wisdom would have us not stock the item and take the downtime hit. Of course, we may go out of business while we wait for the part

How to calculate a safety amount of stock:

The safety stock is the difference between your average usage of a part and the maximum probable usage of that part for a period equal to the lead-time.

Safe = (Um - Ua) * L,where

Um = The maximum probable usage for that part per week
Ua = The average or usual usage for that part per week
L = The lead-time in weeks

Example: The safety stock for a part where 12 is the average weekly usage, 18 is the peak weekly usage, and a the lead-time is four weeks would be (18 - 12) * 4 = 24

How to calculate the Minimum or Reorder Point (ROP):

The minimum stock level or ROP is the amount of stock that would trigger a reorder. In theory, your stock level should use up the stock and fall to the safety stock level on the day the truck brings your reorder. In a JIT shop the safety stock is zero.

ROP or Min = (U * L) + S where

U = Average usage of the part per week
L = Lead-time for that part in weeks
S = Safety stock

Example: Using the same example the average usage is 12 per week and the lead-time is 4 weeks. We calculated the safety stock at 24 parts: (12 * 4) + 24 = 72

How to calculate the E.O.Q. (Economical order quantity):

The economical order quantity is the number of parts to place on order at one time. This area is where the thought is changing, with consideration of the use of blanket orders to lower processing costs. This formula gets significantly more complex when the unit price varies with the order size (as it frequently does). If the unit cost varies significantly, the formula has to be run at each price break and evaluated. JIT attitudes have a major impact on the E.O.Q. With JIT deliveries you would work on reducing the `C` factor and increasing the carrying costs `A`, to reduce the E.O.Q.

$EOQ = \sqrt{(2\ (CN)/(UI+A))}$ where:

C = Cost of processing an order, receiving and stocking the material. Frequently C is in the $50-$100 range but might be a lot higher for advertised bids. This cost may also be called COA or cost of acquisition

N = Yearly usage of the part

U = Unit cost of the part

I = Rate of interest sometimes called the opportunity cost.
'I' could be the prime lending rate that banks charge their best customers
or a higher opportunity cost (how much return could the company get if it invested the money in itself?).
Some formulas for EOQ add in a factor to `I` for the cost of a stock-out.

A = Annual carrying cost in the store room, usually in the 10-20% range

Example: Our part has an annual usage of (12 * 52) = 624 units per year. The unit cost is $10.00. We will use $100 to process the order, 10% for interest, and 15% carrying cost.

(2 (100 * 624)/ (10 * 0.1) + 0.15)
=124,800/1.15 = √ 108,521 = 329 is the E.O.Q.

How to calculate turns:

Numbers of turns evaluate the efficiency of the money invested in the maintenance inventory. When turns equals 1 the inventory value is used up once a year.

Turns = (P/S) Where:

P = Annual dollar purchases of parts and spares
S = Stock level in dollars

Example: If your department purchases $1,200,000 each year and the inventory on the shelf equals $600,000 then Turns = (1,200,000 / $600,000) = 2

Saving money by reducing inventory on the shelf:

Big ticket analysis - Most of the dollars on your shelf are tied up in a relatively few big ticket items. These $500 to $50,000 and up parts represent more than 50% and as much as 80% of the dollars invested. Steps to free up money and space, start with making a list of all of the big ticket items. Ask four questions about each part:

1. Does this item belong in the inventory? Has the unit for which it was pur chased been retired?
2. Is the item a special insurance policy item (long lead-time part for a critical machine)?
3. How many should we really stock?
4. Is there an alternative strategy for handling this part. Such strategies might include vendor stocking, consignment stock, group stocking (one plant holding the part for several potential users), re-engineering to eliminate the need, or taking the risk of not having the part and waiting for it.

To reduce your inventory, look at the answers to these four questions. In most inventories, 10% to 20% of the dollars can be eliminated without impacting the ability to respond to maintenance demands.

Traffic

The mission of traffic is to move freight into and out of the plant by the most economic method consistent with the delivery timing requirements. Traffic saves the organization money by knowing the best modes, carriers, and strategies for moving freight. Maintenance usually represents a minor headache to the traffic department. Typical maintenance shipments are LTL (less than truckload) and time sensitive. Modes of shipment include truck, rail, air, and other. Within each mode there are several distinctions.

Truck can include small package delivery, LTL, truckload, and special trailers (such as drop frame, flatbed, refrigerated). Truck also includes permitted loads (too large, heavy, or dangerous for conventional moves). Rail cars have similar characteristics but have almost double the size and weight capacity of trucks. Types of rail cars include boxcars, tank, flat, and special cars.

Airfreight includes small package delivery service, regular air freight, regular airlines airport to airport, and charter. Occasionally, helicopters are used to deliver freight to places that are hard to reach. Other shipment modes include all special moves such as barge, ship, courier, and anything not already mentioned.

Costs within each mode are determined by weight, distance, and class of commodity. For example, one of the lowest cost commodity classes is Class #50 Iron and Steel fittings. The reasons that class 50 is inexpensive to move by truck are that the products are hard to damage and dense (small cubes with high weight) and not usually desirable for theft. A high-cost commodity class would be computers, which are easy to damage and are of low density (the trailer would be full before weight limits are exceeded). Theft exposure is also high.

Terms used in shipments:

FOB (City, Shipping point or Delivered): Free On Board means that the seller will load the truck or rail car and pay for loading the freight but not unloading. The FOB point is important because of both the responsibility for the shipment and the freight charges. FOB delivered keeps the vendor responsible for the shipment until it reaches your door. FOB shipping point or FOB originating city (also called FOB Ex works) makes you responsible for the shipment. If an FOB Philadelphia shipment is damaged, you still have to pay for it, and then submit a claim to the carrier. If the freight is shipped FOB Delivered, the damage is the shipper's problem. In all examples, you pay for unloading.

FAS: Free Alongside (you are responsible for the loading charge) commonly used for ships or very large freight.

For most maintenance departments, the preferred method is UPS ground, or FedEx ground, on small shipments. If more speed is needed, use Federal Express Economy or UPS second day, then Federal Express Priority or UPS Red label. Large shipments should be handled by other airfreight companies such as Emery, DHL, etc .

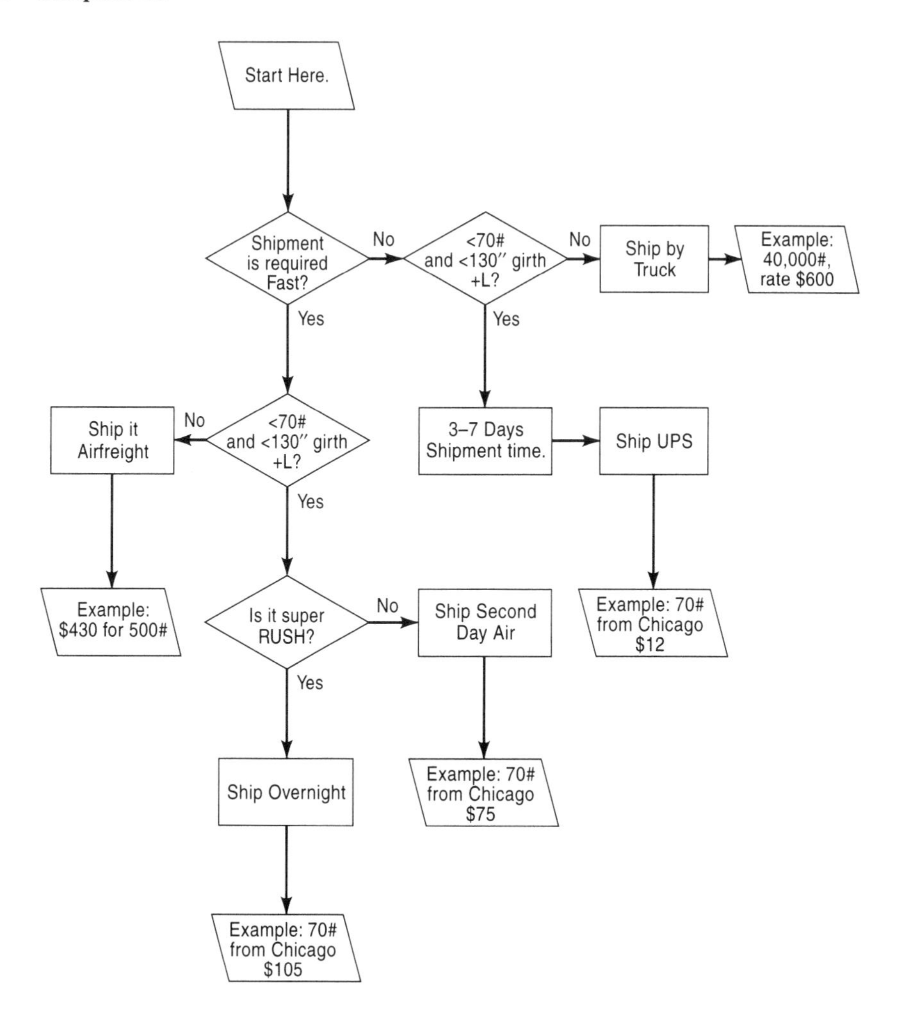
Start Here.
Shipment is required Fast?
No
<70# and <130″ girth +L?
No
Ship by Truck
Example: 40,000#, rate $600
Yes
Yes
3–7 Days Shipment time.
Ship UPS
Example: 70# from Chicago $12
Ship it Airfreight
No
<70# and <130″ girth +L?
Yes
Example: $430 for 500#
Is it super RUSH?
No
Ship Second Day Air
Example: 70# from Chicago $75
Yes
Ship Overnight
Example: 70# from Chicago $105

Figure 49

Zero-Based Maintenance Budgeting

There is tremendous pressure on maintenance managers to improve their budget performance.

Traditional budget methods do not seem to be effective in the maintenance arena because maintenance expenditures are made up of thousands of seemingly unrelated events. Maintenance does not seem to be volume related (higher output equals higher maintenance). The breakdowns and other maintenance activities are hard to predict and do not necessarily relate to what broke last year. To successfully budget (and therefore predict) maintenance expenditures, we must divide the whole maintenance demand into its basic parts.

A zero-based budget breaks the overall demand for maintenance services into its constituents, that is, assets or areas. Look at each asset (or group of like assets) to determine the maintenance needs. In addition to the unit or asset list, a zero-based budget has allocations for certain areas that are hard to define as individual assets, such as the electrical distribution system, or the paved parking area and sidewalks.

Prior computerization of maintenance simplifies the construction of a zero-based budget. The computer can easily generate an asset and areas list. Many systems allow you to create classes of equipment where like equipment is aggregated into one line. If the system has been in use for more then a year, you can attach the hours and material dollars for each asset and area. Some systems have space for a reason for repair (see chapter on work orders). The reason for repair would roughly correspond to the categories below. Most systems allow export of the files to a spreadsheet for further manipulation.

All maintenance activity can be traced back to one of the eight demands that follow. Shops that are craft dominated have a more complicated problem. After the budget is completed, people must go back to the individual demands and break out the labor by craft.

The eight reasons for having maintenance resources are:

1. PM/PdM- preventive maintenance and other routine work expressed in hours/materials. Based on your facility and equipment size, use, construction, and

the standard times of the PM activities, you can predict how much time and materials PM's will take. In a TPM shop, some of the PM hours will come from operator's records. The simplest formula is to multiply the number of services by the time for each service. Also, look at the materials used for each service. Include some time for the short repairs that the mechanic will get done during the PM. You have some flexibility in scheduling so you can consider PM's as a level demand throughout the year.

Specifically, PM work includes all the inspection, adjustment, bolt tightening, oiling, cleaning, and readings that are called out on task lists. The task lists are constructed on a periodic (quarterly, annual) basis. Routine work might include Monday morning line start-up, daily checks, and other routine jobs that are done periodically, where the work content is known and doesn't change.

2. **CM**-corrective maintenance hours/materials, also called scheduled repairs or planned maintenance. As your PM inspectors examine each part of the facility and all the equipment they write-up problems that they see. These write-ups become the backlog of corrective maintenance (CM) for the maintenance schedule. The repairs are planned, and are chosen for scheduling in the weekly coordination meeting.

Where to put short repairs is an important question. If your CMMS permits it, CM would include short repairs. Short repairs are jobs done by the PM person when a unit is repaired during the PM. The short repair is added as an extra line or task to the work order but it should be coded to CM rather than PM. If the CMMS does not allow added tasks, the short repair would be charged to the PM but noted in the comments of that repair order.

You can look at previous years to get an idea of the hours for these activities. You have control of the schedule so this demand can be considered level throughout the year. These scheduled repair hours are inserted by equipment, by group of like equipment, or by area.

A world class manufacturer might have 55% of their workload in this category. CM should at least be the largest category of work. Some CMMS report this category with the PM hours and call it all PM. CM is the work load for the backlog relief crew.

3. **BR**- This category includes all types of breakdowns from the routine broken pulley on a conveyor to a $1,000,000 catastrophic forced outage. Without inspection and inspectors the operators find problems first. Operators or others tasked with inspection such as safety inspectors, also are the first to find vandalism, safety issues, breakdowns, and damage. At the beginning of the year, budget the same

amount of hours for B as the previous year by asset or category. If you've made changes to PM/PdM you can slowly reduce the amount of B hours.

At the end of the year you can back off from B as the PM system starts to take effect (assuming you've made changes). For purposes of budgeting, B creates a level demand, but emergencies will tend to bunch up. Factories use overtime, outside contractors, or ignore problems for a while to level the demand for B. In larger facilities with multiple crews, this work will look more and more level. See seasonal demands (SM) for a special example of UM demand.

4. **UM**- all types of user maintenance (hours/materials) applies to all requests from users/customers. This work extends from routine line start-up tasks, through small informal projects, to large installations. You can design the sub-categories to suit your plant. For example, the work might include UM-R (Routine work), and UM-P (Small projects), etc. UM includes servicing minor user requests for hanging pictures, moving furniture, and other personal service.

Responsiveness to user requests is essential if maintenance is to be viewed as effective. This statement applies to B and UM categories. Most users will judge you entirely on how you respond to their requests (other benchmarks usually don't have as much impact on their quality of life).

5. **SM**- Seasonal Maintenance hours/materials. This category includes all special seasonal demands. Your entire grounds maintenance effort is certainly driven by season. Inspection of roofing systems before summer and winter, and checking air conditioning before summer, are seasonal demands.

Some businesses are seasonal. Cleaning the Candy Cane line before it starts up in July would be a seasonal demand. You can also use this category to pick up some percentage of the seasonally-driven emergencies or seasonally-driven PM. Budget hours at the beginning of each season by asset or group based on history.

6. **RM**- Replacement/Rehabilitation/Remodel Maintenance hours/materials. In some organizations, this category is called capital improvement and is handled outside the normal maintenance budget.

RM also includes all maintenance improvements and efficiency improvements. At some point, units that have not been maintained for a period of time, or have reached the end of their useful life, will have to be rebuilt or replaced. The rebuilding effort should be added to your maintenance budget as a capital replacement line item, separate from any current maintenance activity. If your people are doing the modernization to bring units up to PM standards, the hours will have to be budgeted.

You have control of the rebuild schedule so you may be able to use rebuilds as a crew balancing tool. A special case of RM is Management decision. This work is called for by a manager when the decision is made to change something in, on, or around a machine, other asset, or the building. The reasons for the decision might range from output improvements, improvements to reduce quality variations, energy efficiency, improve usage, legal problems, or even a whim (I hate yellow presses, paint them green!).

Maintenance demands for the whole operation (not tracked by individual but by location). After the basic demand has been cataloged by equipment or area of the plant, look into some of the budget busters below. A well designed budget can be ruined by excessive social demands generated by visiting dignitaries, or a large construction project's effect on the remainder of your operation.

7. **SD-** Social Demands (sometimes known as hidden demands because they don't always show up on work orders). This category also contains elements of PS (Personal Service). Your primary mission is maintenance of the equipment and facility. You may be called upon for other duties in your organization. These duties may include supplying clean-up people, running tours, preparation for visiting dignitaries, providing chauffeur services, picking-up or delivery of papers or packages, organizing picnics, or work on non-organization equipment and facilities (charity work). Estimate your hours for these activities. Tip: Create work orders for all this kind of work for some fun year-end reports.

8. **ED-** Expansion demands. Any expansion in the size of your facility such as number of lines, size of your work force, additions to the scope of your control, will add hours to your overall requirements. New buildings, assembly lines, and major changes to the plant require significant maintenance time (even if they are done entirely by contractors). New facilities disrupt current activities as well as taking direct time. Adding satellite facilities will result in additional lost time until systems are well in place. Estimate additional time if an expansion is contemplated.

9. **CD-** Catastrophic Demands. Every location seems to have characteristic catastrophes. Add time for one or two catastrophes. Review your records for the amount of time spent in a typical catastrophe. This time can include floods, blizzards, hurricanes, trucks taking out the side of the buildings, fires, etc. It might seem funny to try to budget for catastrophes but in some plants they are not rare events. While they cannot be predicted in specifics they can be looked at in general.

How to Set-up a Maintenance Budget

1. Start the budget process by compiling a list of all the machinery and equipment that you maintain. As much as possible, arrange the list by department or cost center to facilitate report printing at a later stage. If you have a CMMS, (Computerized Maintenance Management System), print an asset or equipment list. This list might have as few as hundreds, thousands, or more entries depending on the size of your plant.

2. Add to this list, areas of the plant and site that require the kind of maintenance resources that don't lend themselves to the unit concept. Typical areas include roofs, pavement, electrical distribution systems, piping, doors/windows, etc.

3. Examine the list to see if there are any units that can logically be grouped together. A wire harness assembly plant might have 50 braiding machines of similar usage and vintage. For purposes of a Zero-base budget these machines could logically be aggregated into one line. Putting similar units or areas together simplifies the process and makes predictions more accurate.

4. Collect any maintenance data available by unit or area for the last several years if available. Your CMMS would facilitate this step. If the data is coming from the CMMS, see if it has an export capability. Some systems will send data to spreadsheet files without re-entry. Inquire if your accounting or cost accounting group can provide details of the costs to maintain certain areas, departments, assets, or production lines, with a view to reducing them.

5. We recommend this whole mass of information be designed into a computer spreadsheet. Create a template to duplicate the form at the end of this section. The equipment, areas, and groups of units/areas are listed in the template.

6. After listing the individual units and the general assets, add the global lines (that apply to the whole site) social, expansion, catastrophes. Look into your history or estimate the impact of these areas. The three areas can be added as hours and materials or as percentages depending on the need. If these areas have traditionally been non-work order items, now would be a good time to set up the codes to put them on work orders. Once accounted for, these costs can be studied year to year.

7. Once assets have been inserted into it, the template becomes the basis for the zero-based budget. Back-up the filled-in template on a separate drive. You have put in many hours at this point, so make your back-ups now and keep them up to date! This computerized list might have multiple uses, so spare copies might be useful for other reasons.

8. Review each unit, area, or group, and estimate your PM, CM, B, UM, RM, and SM costs and hours. A useable history of costs from accounting or from the CMMS greatly simplifies this process.

9. Add in your estimates for SD-Social, ED-Expansion and CD-Catastrophe related demands against the department. These values can be entered as percentages of the above areas, or as hours and material costs.

10. Your material costs are the sum of all the material columns, your hours are the sum of all the hour columns. You would then apply the costs of your labor, fringe benefits, and maintenance overheads to determine your budget.

When management wants reductions in your budget you have a new level of discussion. All changes need to be justified in terms of higher or lower levels of service on individual assets or areas. When cuts are needed, you can talk about which assets will be allowed to deteriorate or which departments will not be served as well. Almost every business has deferred maintenance. You may see a problem slowly developing and put off the work. You could be short of funds, be planning a major rehabilitation, planning to sell the unit or property, or lack the requisite skills. Some organizations run their whole operation with excessive amounts of deferred maintenance. Distribute your zero-based budget to the users, staff, and top management, for comments.

If your current hours available are only a small percentage of your budgeted demand, then some things will not get done. Either deterioration is taking place, or your customers will be unsatisfied, or both. One solution is to use contractors to make up the short fall. Some organizations are using this strategy to maintain maximum flexibility.

Using the budget to schedule the need for supplementing the crew with contracting

Some organizations use outsourcing strategies where they provide crews for 75% to 80% of the demand and use outside vendors during peak periods. The most effective way to predict the need for contract labor is to recast the budget on a monthly basis. Using the hours per month, you can see which months will exceed your crew available hours.

The process is similar to a staffing exercise. If your core crew has 1400 hours available a month, the contractor would have to supply any labor above 1400 hours. The budget will show the months in which it is likely that contracting will be needed. Moving project work dates can minimize contractor needs in a given month.

ZERO-BASE MAINTENANCE BUDGET Organization: Department: Name: Page___of___

ASSET or GROUP	PM Hours	PM Mat'l	CM Hours	CM Mat'l	UM Hours	UM Mat'l	SM Hours	SM Mat'l	RM Hours	RM Mat'l	TOTAL HOURS	TOTAL MAT'L
Sub-TOTAL												
SOCIAL Costs												
EXPANSION												
CATASTR-OPHE												
GRAND TOTAL												

Figure 50

Shutdowns, Outages, and Project Management

On larger jobs the techniques of Project Management can be used to control the project to good effect. Note that project management was one of the first areas to be transferred to both big computers and microcomputers so there is now a wide variety of software packages at all price ranges.

There are techniques to manage large repair projects efficiently. The planner should be conversant with the major ways to set up and manage jobs. In a nutshell, the Project Management packages require you to list all the sub-projects, resources, and dependencies. From this information, the software builds a model of the project and tracks it in real time, producing alerts when you fall behind, or when there are resource conflicts.

Project Management	Maintenance Shutdown Management
❑ Many related jobs ❑ Logical steps interrelated to an end ❑ Scope of work is usually pretty clear and does not change ❑ Project is organized around cost codes, and a hierarchical job structure ❑ Span of time is weeks and months ❑ Schedule can be updated on a weekly or even monthly basis ❑ Understandable end point (building is complete) ❑ Less need for safety permits and clearances ❑ Can be planned well in advance ❑ Software, big issue is critical path. ❑ Staffing requirements are more static ❑ Span of time is weeks and months	❑ Many unrelated jobs (1 per work order) ❑ More unknowns and greater emergent work. ❑ Many one-step activities ❑ Span of time is measured in hours and shifts ❑ Scope of work is not always clear and is undefined until the beginning of the outdown ❑ Big issue is resource leveling ❑ Scope may change as items are disassembled. ❑ Much of the work is invisible (inside tanks, rebuilding pumps, etc. ❑ Planning must wait until the scope is pinned down, which is later in the process ❑ Updating must be by shift or even more often ❑ Extensive safety permitting required ❑ Staffing levels can vary widely ❑ No easy end point

Three Major Costs

There are three major cost items in a factory or process plant.

1. Capital investment including building and extending or expanding a plant or facility. These projects might be run while the rest of the plant is still producing.

2. Catastrophic breakdowns, forced shutdowns, and large accidents, fires, and explosions.

3. Intentional scheduled shutdowns, outages, and turn-arounds.

The techniques of shutdown management are designed for item three. Project management (related to shutdown management) is best suited for item #1 if there is no existing plant. Shutdown management is used where the project will be positioned close to an existing and running plant.

Size

Shutdowns come in a variety of sizes. Large events require dedicated management teams, fancy software, and dedication of possibly thousands of people. Small events are closely related to large maintenance jobs.

Seven factors to successful shutdowns (partially adapted from Managing Maintenance Shutdowns and Outages by the author)

Organization: The best organizations assign a manager who has responsibility and authority, planners, support personnel, people who have the time, and all preparation skill sets covered for an adequate period.

Planning and scheduling: Thinking through the jobs, anticipating problems, developing plans and contingencies for when something goes wrong. How to perform the work. Defined overall scope, work lists, control, prefabrication, design of schedule, keeping on schedule.

Contractors: How to integrate external organizations. Create accurate contractor packages, identify and evaluate contractors, identify sub-contractors, build in carrots and sticks, provide mobilization plans to ensure the right people are there on day one, demobilization plans to manage unnecessary costs.

Accounting, Costs: How much did the shutdown cost? How to estimate, report, and control costs. How to fund, estimate, refine, develop contingency, cost reporting systems in real time, close out. Evaluate financial risk.

Logistics: Organization for the parts, materials, and supplies. Elements of logistics include a site plan, site control, control, safe routes for lifts, off-site management. Functions are parts receiving, storage, and job site delivery.

Execution: How to manage in the face of reality. Management control, pre-start briefing, daily routines, shutdown of existing plant, work the plan, deal with whatever comes up; keep your eye on the ball until the game is over. Dealing with risks and developing work arounds for risks that were unanticipated.

Reporting: What happened? Were the lessons learned preserved? Are customers satisfied with how they were kept up to date? Can we avoid making the same mistakes again?

Metrics: One of the most important tasks for shutdown management is to have a set of goals to measure the success of the shutdown. These goals help express top management's desires and expectations. The goals would have values in several areas.

- ❑ Shutdown came in on or under budget.
- ❑ The plant was put back on line on time, or early.
- ❑ The plant started up smoothly.
- ❑ All jobs were completed.
- ❑ There were no lost-time accidents.
- ❑ There were no unscheduled discharges of material into the air, water,or on to the land.

History of project management

One of the greatest areas that was developed in the 20th century is project management. The field of modern project management really started in 1914. We are indebted to Harry L. Gannt of the Frankford Arsenal in Philadelphia for developing a systematic technique for tracking and scheduling projects. Developed in 1914, the Gannt chart is one of the oldest planning tools available to maintenance managers.

Since then, dramatic improvements have been developed to manage larger and more complex projects. The critical path method (CPM) developed by Remington Rand and DuPont, improved the ability to determine he critical events that could hold up a large project. The Project Evaluation and Review Technique (PERT) developed by Lockheed and the consultants Booze, Allen, and Hamilton developed at about the same time as CPM.PERT included most optimistic, most probable, and most pessimistic, and was adopted by the military.

One of the first such projects on which PERT was tested was the design and assembly of the first nuclear submarine (Polaris class) George Washington. The Navy and Electric Boat Co. in Groton, Connecticut had to schedule 250,000 major activities of 250 contractors and 9000 sub-contractors in the multi-year project. At any given time, delay in any one of hundreds of activities could throw the overall project off schedule.

The PERT charting and management method was used to plan this project. In the beginning they were not used to manage the projects because that required too much computer resource. Because of its complexity, the difficult task of maintaining the critical path with the project changing, and with activities being complete on schedule, ahead of schedule, and behind schedule, the entire system was programmed and run on the then powerful IBM mainframe computer. The project was completed 2 years ahead of schedule and under budget.

The first use of CPM was in construction of a DuPont catalyst plant. DuPont went on to use CPM for the first shutdown of one of their chemical plants in Kentucky. The duration of the shutdown was projected at 125 hours. With CPM analysis they completed the project in only 97 hours.

These planning concepts are extremely powerful:

1) There exists a group of activities within the project, the sum of whose times regulates the length of the project (CPM and PERT). The longest path through the project (that includes these activities) is called the critical path.

2) We also know that time estimates are more likely to err by being too short rather then too long (PERT only). On the Polaris program the engineers used an approach called the Beta distribution which is not symmetrical. The more pessimistic (overdue) estimate had the greater probability.

3) If you keep the ever-changing critical path on schedule, the project will run on schedule.

4) Conversely, if the critical path falls behind schedule early in the project you know:

- a The whole project is in trouble
- b Only an intervention (more labor, expedited material deliveries, etc.) can bring the project back on track.

5) Finally, collisions of Labor/Material/Tooling/Machine/Order are substantially easier, cheaper, and faster to resolve on paper than in the field. This maxim is also a golden rule of planning.

How to set up a Project Management Chart

1) List all activities for the project

- a An activity has a defined beginning and ending
- b No other activity has to start in the middle of the activity. If one does then the one activity should be split into two activities.
- c Determine the immediate successor activities for each activity
- d Determine duration for each activity
- e Determine effort level (size of crew)
- f Determine other resources for each activity

2) Start adding activities to the PMS and all the information available.

3) Be alert for conflicts in Labor/ Materials/ Tooling/ Machine/ Safety/Permits/ Sequence. If you chose an added effort level, then you can have the software add the columns by craft to determine the numbers of people required for the project.

4) All the activities along the longest path are critical path activities. Slippage in any critical path activity will result in the project being late.

5) There is a certain point that, if delayed, non-critical path activities become critical.

6) After the project is started, mark off the activities that are completed. Estimate the durations of the uncompleted projects and shift them on the chart. Recalculate the critical path. If anything is out of bounds then plan your intervention.

19

Personal and Personnel Development

Craft Training

The skills needed to run today's factories and buildings are changing faster than people can adjust. Technology jumps disorient even the most dedicated workers.

Case in point: Small factory manufacturing fuel control systems: Chief of field service, Calvin Smith was 55 years old. The workforce consisted of three younger technicians. Smith was required to do service himself and usually took (or was brought) the most challenging problems. He was the best and most highly-skilled trouble-shooter for over 12 years through the transistor era (he actually started with relay logic).

The company moved to CMOS integrated circuits. Now each integrated chip replaced an entire circuit board. After a painful learning process, Smith came up to speed on CMOS. He never developed the comfort level with integrated circuits that he had with transistors. Instead of knowing and following the entire logic of the board he associated certain chips with certain faults and replaced the chips until the board worked. Toward the end of two years his expertise was quite good. His attitude had recovered.

After two years of CMOS, new microprocessors first started to show up in the designs. Smith started over, but the difference between microprocessors and CMOS was much wider than the difference between CMOS and transistors. He never understood the concept of programming, or the dynamic nature of the data and address bus. He couldn't understand how to trouble shoot a dynamic system such as a typical microprocessor board.

Our dedicated field service manager took very early retirement feeling that the world had passed him by. The company lost his expertise. He lost his sense of mastery and feeling that he was part of an important field. He now does odd jobs (he started as an electrician) in his neighborhood. Interestingly, his old subordinates still bring him problems, at which he throws himself with relish.

This event was a waste of human resources. Proper training would have saved this person. Although the company spent money to train the engineering staff it didn't think to include the service staff. The design lag time (it took two years to get the first microprocessor product off the drawing board) would have been more than adequate to train the entire staff.

Attitude, Aptitude, Ignorance

Caveat: Before embarking on an elaborate training program, ask yourself the question, what is the cause of lack of performance? There are three general causes: attitude problems, aptitude deficiencies, and ignorance or lack of knowledge.

At the core of the question is; does this person need training, some kind of counseling, or are he or she unsuited to do the job? Some performance problems come from a bad personal attitude. Many more performance problems stem from either inadequate training or lack of ability such as strength, reach, or intelligence (which cannot be overcome even with practice and training).

To make the diagnosis decision more difficult, many people develop attitude problems as a defense mechanism against feelings of ignorance or incompetence.

Types of competencies

In training terms there are three types of learning that apply to this situation; Knowledge, Skill, and Attitude. For higher-level jobs (such as chief service person) all three must shift to competence. Many types of training address one or other of these types of learning without regard for the other. Maximum effectiveness in this instance must encompass all three areas.

TYPE	OBSERVABLE BEHAVIOR	PERFORMANCE LEVEL
1. Knowledge	be able to describe diagram, argue, etc.	answer X of 10 questions correctly
2. Skills	demonstrate, show, perform, solve	do. . . in X minutes with no mistakes
3. Attitude	comfort, without hesitation	to your own satisfaction

Figure 52 Competencies

Let's analyze Calvin Smith's situation from a training perspective.

Calvin needs competencies in all three areas to be a good chief technician. He lacks certain knowledge that could show him what is happening on the board. He lacks some specific skills to help him fix the boards. His biggest problem now is a negative mental attitude. His attitude stands in the way of his gathering the skills and knowledge. The attitude will be the last to be fixed, and will follow mastery of the skills and knowledge.

There are four steps in the design of tailored learning program:

Step 1: Determine what knowledge, skills, and attitudes are needed for the job. Before we can look into teaching anything, we have to see what is needed. Look at the job as it is today, and forecast where the job is going in the short term. The big picture of competencies is called the General Learning Objective (GLO). The concrete and specific skills, knowledge, and attitudes required to do the job are called specific learning objectives or SLOs. If properly designed, a person achieving these SLOs would be successful in this job.

What is needed to trouble shoot microprocessor boards? We must decide what level of competence is appropriate for a service chief in servicing these boards. Examine form 1. This form should be kept with the job descriptions for the Chief's job for future reference. Note that we are discussing only the microprocessor part of the job. The chief's job has many other facets.

Step 2: Evaluate the potential trainee's (or trainee groups') current skills, knowledge, and attitudes. A direct supervisor might be able to make an educated guess. If the trainee have good insight, they might know where they are weak. See form 2. Most situations require some kind of testing (either observation on the job, or more formal written or bench tests). The testing should be designed to uncover the skills on your required list from step 1 (form 1).

It is important to note that success on the test should correspond to success on the job. Testing that does not reflect job requirements is said to be invalid. In the US the Americans with Disabilities Act (ADA) and related legislation state clearly that the test must not discriminate against any group, disability, or condition. For example, if the worker must lift 100 pounds in the test, the job must call for heavy lifts where equipment cannot easily be used.

We must evaluate our service manager's skills, knowledge, and attitudes. After we make a list of specifics we rate him/her (or test) in each area. The result would be a test tailored to one individual. Form 2 would be kept in a personnel or training file for the individual.

Step 3: Translate the voids in skills, knowledge, and attitudes of the potential trainee from the required list to develop a training lesson plan. The training plan should list all the types of learning that this person/group needs. Form 3 summarizes the skills, attitudes, and knowledge that Calvin Smith lacks and needs for the job. The form also recommends possible exercises and resources to provide the learning that Calvin needs. Form 3 also estimates the time requirement for the trainee and the requirement for any supporting staff.

Step 4: Did the candidate learn all the skills and knowledge and adopt the attitudes needed from the training to be successful on the job? In simplified language, was the training successful? If it wasn't, then the candidate must be retrained.

You might go through this exercise for all related jobs. The service technician might have related SLOs that can be incorporated economically into this training.

The question of Return on Investment

Analysis is complete except for the important go/no-go question. Is this training worth the investment? Is the investment of about $8700 worth the probable returns? The returns come in two areas:

What is Calvin Smith worth as a chief technician? He has 8-10 years left that would be spent with our company (we presume, but cannot guarantee). We have no guarantee that this program will work or that Calvin will learn what he has to know to feel successful. Does his specialized knowledge built up in 12 years with us, and the rest of his experience, have value? Is there a training cost for his replacement? Keep in mind that servicing of microprocessor boards is only a small part of the job of Chief Technician. If we are successful in training Calvin, we might have a good prospect that he might become an instructor for the rest of the field service staff.

We also have a continuing asset in the training program that we have assembled. Let's inventory this program:

One microprocessor training set

All this equipment is packed up in a case ready for use.

- ✓ Chart of normal state of pins
- ✓ Added information on our chip set
- ✓ Additional questions added to test knowledge of use to us. Several boards with known problems in software and hardware
- ✓ Test fixture
- ✓ Program that exercises micro board
- ✓ Boards with common modes of failure
- ✓ Logic diagrams for diagnosis of problems for 20 boards

Figure 53

Form 1

Date 2-23-0X

Dept: Field Service

JOB REQUIREMENT ASSESSMENT

Job evaluated: Chief Service Technician

Candidate evaluated: Calvin Smith.

GLO

GLO (General Learning Objective) followed by SLOs (Specific Learning Objectives), measurement of the SLO included in description. Skills, knowledge, attitudes (noted S, K, A) needed by Chief Technicians: Chief Service technicians should be comfortable and competent in working with all types of faults on microcomputer circuit boards in a field service environment. He/She should also be able to teach the other service technicians.

SLO

1. (S) Use of test equipment, VOM, scope. Must be able to demonstrate competence with each piece of test equipment by choosing the correct tester for the problem, setting it up correctly for the problem, and interpreting the results correctly.

2. (S) Be able to locate specific pins on the chips (power, ground, address, data, reset, enable, etc). Test would include pin identifications on all chips used on board. Pin-out chart allowed

3. (K) Explain what the pins do. Explain and test for normal state of pin such as power, reset, select, address, etc. Trainee should be able to describe the function and normal states of every pin on all chips used on boards. Use meter or scope to verify knowledge

4. (K) Know basic chip types and their uses: Micro, RAM, refresh circuits, ROM, PIA's, specialized drivers, etc. Describe function of all chips on board in normal operation.

5. (K) Be able to explain the interaction of the different chips on the board. Explain addressing, parallel/serial data movement, timing, etc. Show timing on a clock cycle basis, show addressing and chip select logic, describe data movement via PIA (Sync) to parallel data bus.

6. (S) Be able to write simple programs, in particular programs that exercise the various functions of the board. Write a program that tests all input and output ports, and all memory locations, and reports a defective condition.

7. (S, K) Be able to figure out whether the software has a bug or the hardware has a fault.

8. (S) Be able to set up an exercise fixture, download a program to test and exercise the board.

9. (A) Develop a comfort level with and interest in microprocessors.

10. (K, S) Develop logic for dealing with common problems (reset held down, address line grounded, RAM chip bad, etc.)

11. (S) Be able to trouble shoot ten boards with random faults in a day.

12. (A) Feel that he/she can cope with the new technology.

Figure 54 Job Requirement Assessment Form 1 (There would be pages of SLOs for a job as complex as chief technician)

Form 2

Date 2-23-0X

Dept: Field Service

TRAINING NEEDS ASSESSMENT

Job evaluated: Chief Service Technician

Candidate evaluated: Calvin Smith

(GLO) Skills, knowledge, attitudes (noted S, K, A) needed by Chief Technician to trouble shoot microprocessor boards to a component level (Only SLOs where a deficiency is found are to be included)

Evaluate trainee's competence in the SLOs determined to be job requirement SLOs

SLO	Competence
1. (S) Use of test equipment, VOM, scope	Demonstrated knowledge of test equipment
2. (S) Be able to locate specific pins on the chips (power, note that memorizing pin locations is not an objective	Can locate specific pins if given a pin-out chart, ground, address, data, reset, enable, etc).
3. (K) Explain what the pins do. Explain and test for normal state of pin such as power,of reset,select, address, etc.	Needs some work on identifying normal state and sig natureaddress, data pins
4. (K) Know basic chip types and their uses: Micro, RAM, refresh circuits, ROM, PIA's, specialized drivers, etc	Knows some chips (from earlier work) needs training in others.
5. (K) Be able to explain the interaction of the different chips on the board. Explain addressing, data movement, parallel/serial timing, etc	Needs some help here; can explain some of the interactions suchas chip select, reset.
6. (S) Be able to write simple programs, in particular programs that exercise the various functions of the board.	Needs help here
7. (S, K) Be able to figure out whether the software has a bug or the hardware has a fault.	Needs a lot of help here
8. (S) Be able to set-up an exercise fixture, and down load a program to test and download a program exercise the board.	Has built many test fixtures, needs to be shown how to
9. (A) Develop a comfort level with and interest in microprocessors.	Needs help here
10. (K, S) Develop logic for dealing with common problems (reset held down, address line grounded, RAM chip bad, etc.)	Needs some help, has done this for other technologies
11. (S) Be able to trouble shoot ten boards with random	Needs help here faults in a day.
12. (A) Feel that they can cope with the new technology.	Needs help here

Figure 55 Form 2 Training Needs Assessment

Form 3

Date 2-23-0X	Dept: Field Service
TRAINING AGENDA	Job evaluated: Chief Service Technician
Candidate evaluated: Calvin Smith	

1. GLO) Skills, knowledge, attitudes (noted S, K, A) needed by Chief Technician to trouble shoot microprocessor boards to a component level (Only SLOs where a deficiency is found are to be included)

2. Preliminary Cost Estimate:

Student (100 hr.) at $35	$3500	
Engineer (71 hr.) at $45	$3195	Time Estimate: One hour every other morning for 6-10 months, plus several full days to build fixtures and test boards
Outside Course	$ 410	Engineer: Preparation 45 hours, Review 1 hour/week
3 day seminar afterward	$ 995	
Total cost	$8700.	

3. (K) Explain what the pins do. Explain and test for normal state of pin such as power, reset, select, address, etc	Purchase Heath Kit's Micro-Processor Lab (be sure the lab uses the same processor as we do. If not, find a similar kit that uses our processor). Review course that includes training in the normal states of pins. Have engineers add an exercise to make Calvin plot the normal states of all important pins on one of fy the correct ness of his chart. Keep this chart as part of the training program
4. (K) Know basic chip types and their uses: Micro, RAM, refresh circuits, ROM, PIA's, specialized drivers, etc.	Add information for any chip that we use not covered in course. Design short test for our chip set.
5. (K) Be able to explain the interaction of the different chips on the board. Explain addressing, parallel/serial	Review tests at the end of the chapter to see if Calvin understands the information. Add questions in areas not well covered data movement, timing, etc.
6. (S) Be able to write simple programs, in particular programs that exercise the various functions of the board.	Course includes several programming exercises. Add one exercise. Exercise: design a program that tests all the functions of our board
7. (S, K) Be able to figure out whether the software or hardware has a fault.	Engineer is to introduce known problems into several boards including hardware faults and software bugs. Calvin to build a test fixture and work through logic of diagnosis to the satisfaction of engineer and Calvin. These boards, fixture and logic notes should be kept as part of the training program
8. (S) Be able to download a program to test and exercise the board.	Will be covered by engineer in #7
9. (A) Develop a comfort level with and interest in We microprocessors. Microprocessor	Ask Calvin, from time to time, how his comfort level is with the new technology. expect it to increase as he gains mastery . Get Calvin a subscription to process control magazine. Ask him to report on new developments
10. (K, S) Develop logic for dealing with common problems (reset held down, address line grounded, RAM chip bad, etc.)	Have Calvin review work requests and interview engineers to determine probable failure modes. Create these modes in production boards and analyze symptoms. Add these boards to training kit.
11. (S) Be able to trouble shoot ten boards with random faults in a day.	Obtain 20 failed boards from the field. Repair boards and prepare logic diagram of repair/diagnoses process. Do 3 the first day, 7 the second day and 10 the third day. Review logic for each board with engineer at the end of each day Save logic analysis for training kit.
12. (A) Feel that they can cope with the new technology. to	Question Calvin about attitudes toward new technologies. It might pay at this point wait 3 or 4 months, then send Calvin to an outside 2 or 3 day intensive seminar in microprocessor trouble shooting. He will see how far he has progressed.
13. Last assignment. Work with engineer to organize training materials. Build a storage case and instructions for use. Include the successful logic path to repair all the sample boards that have faults.	This kit will be the nucleus of a kit to be used for training new engineers,service people, and certain production people to maximize the use of the program. For good measure, if space for a desk (bench) is available, have Calvin set up a small training area and design training into all service tech jobs (1 hour a week?)

The Training Function in a Maintenance Department

Training is an essential issue if a maintenance department wants to maintain high quality standards and decent morale. Any organization that doesn't investment in ongoing training is making a grave mistake.

In some areas, training must be a grass roots issue that is handled within the maintenance department or even within an area of the maintenance department. Supervisors might take it upon themselves to begin an ongoing training program by bootlegging resources from other areas, jobs, budgets, or interested parties.

The training function involves the following:

1. Completing a training needs assessment of all maintenance workers. This project involves preparation of several Job Requirement Assessment forms for most of the aspects of the various maintenance jobs, evaluating the maintenance worker's competence in each area, and interviewing the worker for his/her own needs assessment.

2. After the data is collected a training file must be built for each person.

Organizing the training effort

Your department (if not the whole company) should set training goals for all craftspeople. Look at where they are now, where they need to go, and what is missing. The training budget should exceed 1% (20 hours per year) for training, for each person per year. Firms making rapid changes might need 5% (100 hours per year) or more to keep up morale.

The department training point person can be one of the people in the department and the position could be rotated among different people every year. Set up files for each person with the GLO's that they need. Review the GLO's with available time, business cycle, and funds. Be sure to act with the full input of the worker and your supervisor. Once underway, review every file every 6 months to a year and be sure everyone gets an opportunity to be trained.

Some suggestions for non-craft topics are: quality, safety, CPR, fire fighting, toxic material handling, toxic waste regulations, your maintenance information system, statistics, filling out paperwork, PM, scheduling, project management, report writing, shop math, drafting, CAD, computers, engineering, cost accounting, your industry, your end products, and what it's like to operate your machines. The list is endless. Your people will be the better for the attention and the training

Categories for sources of training

Training is big business for a large number of organizations. Before you select a vendor ask yourself some questions.

1. Is this resource the best available to achieve the learning objectives?
2. Is this resource consistent with the style of my department and organization?
3. Will this resource satisfy the expectations of the trainees?
4. Can we afford this resource in terms of both time and out of pocket dollars?
5. Are there other benefits beyond this training from this resource?
6. If it is an outside vendor:
   a. How stable is the organization?
   b. How knowledgeable is the actual trainer about the trade, subject matter, your industry, process, organization, and situation (whichever are important in your setting)?
   c. Will the vendor guarantee results?

Here are some ideas:

1. Your staff is your first choice of potential trainers. 85% of all maintenance skills are learned on the job, so this training is going on every day. Within your staff there are several possible opportunities for trainers. Please note that being a trainer should be viewed as a job-enhancing project. Time should be given for preparation. The trainer should be relieved of other duties for that short period.

✓ Tap your people who are soon to retire as trainers. This group has significant experience that should be passed on to the next generation. In some organizations, people who have already retired are recruited to return as part time teachers. Of course, if the person has bad attitudes, don't let them contaminate the rest of the people.

✓ Use the internal guest instructor concept. In this concept, a staff member would be treated as a guest (given lunch, clerical support, off site premises for longer training sessions, relief from other duties)

✓ Tour training is an excellent team building exercise. Once a month you tour a section of your facility and the most experienced maintenance person plays `show and tell` about the problems and successes in his/her area.

✓ Video technology has rocketed ahead so fast that most firms can afford a quality video camera, recorder, and an editor. New equipment set-up, construction documentation, TPM work, and specific machine training, are popular first subjects. Craft training is a more difficult but rewarding area. After expertise is obtained, any topic can be recorded to good effect.

✓ Look to other parts of the organization such as human resources, data processing, engineering, and production, for expertise useful to your upgrading effort.

2. Many excellent companies provide craft training. These firms provide professional instructors, CBT (computer based training), testing rigs, and video/audio tape. The price, quality, and appropriateness to your operation may vary, so check several vendors.

✓ In-house courses are available on a wide variety of maintenance topics. These courses are most appropriate if many people (7-8 and up) need the same training. Cost is about $1000 to $3000 per day.

✓ Public seminars are very useful for training one to four people. Expect seminars to cost $500 to $2500 and to last 1 to 5 days. Try to get recommendations concerning the better seminars from people in your industry. Most places give discounts for several people from the same site. It is recommended that your people be exposed to others in the field.

✓ Many organizations sell videos. Expect to spend $50 to $500 (some higher) for video training in many areas.

✓ Computer Based Training is a fast growing field. Courses are available for installation on your local computer, to cover the gamut of the maintenance field.

3. Equipment manufacturers. This field of training is also a growing area. Equipment manufacturers have vested interests in a trained user base. Many of them subsidize training, calling it a marketing expense. Time and again it has been shown that trained users are happier users. Negotiate training into all equipment purchase contracts. Excellent, low-cost training is usually available from vendors of predictive maintenance hardware.

Idea for action: Write and ask all your major equipment vendors and major part vendors (bearings, seals, etc.), for any free maintenance, operations videos, or DVDs they have available. Ask for volunteers to screen the videos for usefulness and appropriateness before they are put into circulation.

4. Trade and professional associations. These groups are striving to increase their value to their membership. One of the traditional ways is to provide industry-specific training, in either traveling seminars, or at workshops during trade shows. If your association does not provide training that you believe is needed in your industry, why don't you volunteer to put a seminar package together for the association?

5. Distance Learning. Training delivered either over the Internet or by satellite is called distance learning. These systems train in all areas including electricity/electronics, pneumatics, building trades, business subjects, computer subjects, basic science, and many other areas. You can now get an MBA from a top business school without ever visiting the campus. This mode of learning is another fast-growing area.

6. Technical Schools are an excellent source for trade training. Get to know the people running your local technical schools. Visit and walk through the facility. Many companies donate specific labs, machine tools, benches, or other equipment to the trade school. These donations are of machines or processes that the donor company uses and has people who need training that the technical school can supply. Many technical schools are very open to negotiate training contracts for some or all of your technical training needs.

7. Community Colleges, Colleges, and Universities frequently look for new markets. These institutions have significant expertise in teaching the more advanced subjects to adults. Many of them have entered into instruction contracts with private industry in areas including electronics, computerization, robotics, regulation, automation, business skills, and other areas.

8. Unions are looking at their traditional roles. Many see that skill needs are shifting and have decided to lead the trend by setting up training for their members. Training might be an interesting subject to be discussed if your firm's union is not doing this already.

9. Insurance Companies can cut claims by conducting certain types of training. Some firms will send risk managers through your facility and provide specific training in areas such as safety, risk management, liability reduction, fire safety, storage and handling of chemicals, record keeping for maintenance, safety, and accidents.

10. Governmental Agencies provide seminars and workshops on a wide variety of topics including EPA issues, hazardous materials, waste disposal, safety, record keeping, dealing with overseas vendors, and many others. For example, The Section 8 program under HUD has recently trained all landlords in proper techniques of working with lead-based paint.

There many `rules of thumb' in the teaching of adults. The more rules you follow, the more likely it is that the training will be successful. Larry Davis' leading book on adult training and education Planning, Conducting, and Evaluating Workshops published by University Associates, Inc. San Diego, CA. gives rules for teaching adults that sum up the best thinking on the topic as follows:

1. Adults are people who have a good deal of first-hand experience. Effective training taps into the adult's existing store of experience. The experience of one person in the class is used to train the rest.

2. Adults are people with relatively large bodies, subject to the stress of gravity. Effective training allows the adults to take breaks, move around, and change pace.
3. Adults are people with set habits and strong tastes. Effective training is sensitive to adult habits and tastes, and tries to accommodate as many as possible.

4. Adults have pride. Successful training is careful with the egos of the participants and this care helps develop greater abilities and independence in the areas of the training.

5. Adults are people with things to lose. Good training is concerned with gain and not with proving inadequacy. The most effective training has 100% success ratios.

6. Adults are people who have developed a reflex toward authority. Good trainers know that each adult has a different style of dealing with authority and don't take any reactions personally.

7. Adults are people who have decisions to make and problems to solve. Effective training is problem-solution oriented and entertaining.

8. Adults are people who have a great many preoccupations outside of a particular learning situation. Effective training does not waste the adult's time. Training should achieve a balance between tight presentation and time needed for learning integration.

9. Adults are people with established emotional frameworks consisting of values, attitudes, and tendencies. Training implies change. Change puts a person's framework at risk. Effective training assists adults in making behavior changes. Effective training assists adults in becoming more competent.

10. Adults are people who have developed selective stimuli filters. Effective training is designed to penetrate these filters.

11. Adults are people who are supposed to appear to be in control and who therefore display restricted emotional responses. I Intense training sometimes loosens up these restricted responses. Effective training is prepared for emotional release if it occurs.

12. Adults are people who need a vacation or time off from work. Effective training provides some time away from the grind.

13. Adults are people who have strong feelings about learning situations. Effective training is filled with successes.

14. Adults are people who secretly fear falling behind and being replaced. Effective training allows people to keep pace with the field and grow with confidence

15. Adults are people who can skip certain basics. Effective training starts with where the adult is today and builds on that.

16. Adults are people who more than once find the foundations of their world stripped away. Effective training reminds them of their ability to learn and start again.

17. Adults are people who have ideas to contribute. Effective training leaves room for their contributions.

Common Methods of Training to Consider

Method	Description
Coaching	One on one training and encouragement
Case Method	Analyze a specific incident, problem, situation, or company. Usually done in a group. This method was popularized in business schools like Harvard.
CBT (Computer Based Training)	This mode has all but replaced books for skills training. It includes graphics, animations, voice over, tests, and feedback. Programs can be purchased on CD or down-loaded directly from the Internet.
Distance Learning	Generally learning over the Internet. Comes in several flavors.Can be programmed learning where the whole course is canned and run for each trainee (similar to CBT). Can also be like a real-time correspondence course where the instructor monitors the student's progress.
Distance Learning via satellite	A merger between classroom training with a live instructor and distance learning.The instructor teaches the course to a small group locally while the program is sent up to the satellite. Students that are linked can ask real time questions by E-mail or phone.
Correspondence	Home study of a commercially-produced course of study. An older method but still available. Can be an element of training within a firm.
Demonstration	Trainer shows trainee how to do something. People often can learn a good deal from watching skilled people (think about cooking programs).
Laboratory	Experiments designed to teach by discovery. Learning by doing is the most powerful teacher. There are laboratory-ike courses for hydraulics, electronics, and pneumatics, where the student is asked to trouble shoot a fault or solve a problem.
Lecture	Trainer tells trainee about material to be learned. Today's lectures are frequently accompanied by a computerized slide show (overheads or film slides were used in the past).
Programmed learning	Trainees go through pre-mapped material at their own speed. Can be taken from the Internet, local software, or (in its original form), from a programmed-learning text). Accommodation provided for trainees who need more material in some texts.
Role Play	Trainees play the role and learn from their reactions and the reactions of the other role players.
Simulation	Trainees are presented with a realistic scenario and they work through problems and situations. Control room operators and pilots are trained by this means.

Figure 57

Check list for training:

- ✓ Materials
- ✓ Staff
- ✓ Outside firms needed
- ✓ Trainees
- ✓ Aids
- ✓ Food and refreshments
- ✓ Facilities
- ✓ Accommodations
- ✓ Dates and times
- ✓ Structure
- ✓ Timing
- ✓ Travel
- ✓ Promotion

Supervision and Leadership

What is a great maintenance supervisor?

We discussed the issue of maintenance supervision with hundreds of maintenance managers, maintenance supervisors, maintenance planners, plant engineers, building managers, and production managers, throughout the United States, Canada, Europe, and Asia. The organizations ranged from the largest industrial firms, federal and local governments, to small industrial and building management firms.

There were remarkable similarities in the answers from both giant industrial firms and small firms, and between the federal government and local agencies. The next few pages will discuss the results of this survey.

What are the attributes of a good/great maintenance supervisor?
The answers fell into three general categories:

People skills	Management skills	Technical skills
Good listener	Is organized	Dedicated to quality
Compassionate	Has ability to make decisions	Knows equipment
Can motivate others	Good delegator	Knows job
Fair and consistent	Meets goals of business unit	Knows safety
Respected	Can analyse progress to goal	Can analyze problems
Honest	Knows what is and isn't important	Can evaluate skill level
Effective trainer	Provides good service to customer	Understands product
Open minded	Loyal to organization	
Effective communicator	Oriented toward results	
Coach, not dictator	Good planner	
Good negotiator	High productivity	
Has a cool head	Follows up to see job is done	
Flexible	Understands importance of scheduling	
Can handle pressure	Can assign and keep priorities	
Can read people	Understands and uses budgets	
Adaptable to change	Is available	
Has common sense	Provides a conduit for downward communication	
Willing to learn	Provides intelligence to upper management	
Positive outlook		
An innovator		
Praise in public		
Discipline in private		
Takes control if necessary		
Not afraid to make mistakes		
Treats people as equals		
Can work with different types of people		
Gives recognition for job well done		
Can deal with difficult people issues		

Questions that were asked about Maintenance Supervision

1. What are the attributes of a good/great maintenance supervisor?
2. What is lacking in most supervisors' training (if anything)?
3. If you could give words of wisdom to a new supervisor, what would they be?

Answers to the question: What is lacking in most supervisors' training (if anything)?

- We don't train our supervisors. They learn in the trenches
- Training in listening and communication
- We don't train in how to motivate, how to be a leader, or in psychology (such as how to handle alcoholism, drug abuse, and other employee problems)
- Increased knowledge in allied fields (an electrical supervisor should know some plumbing or pneumatics)
- Training in probable causes, root cause analysis, results of actions/inactions
- How to deal with different types of people
- How to teach skills to others
- Ability to read people and know their limits
- How to supervise friends, older workers, and young workers
- How to deal tactfully with inadequate performance issues
- How to work with budgets
- Increased common sense
- How to keep things impersonal, how not to impose lifestyles
- Administrative training
- Cost analysis and budgeting
- Maintenance management, the big picture
- Knowledge of how best to use the organization's systems and procedures to get things done

Notice that most of the comments from these managers centered on people issues and management issues. Most supervisors who make it in the maintenance department to this level are already technically competent.

We asked people to pass along gems of wisdom to new supervisors:

- Be a good listener.
- Learn to bend, but don't abdicate responsibility.

- ❑ Remember you are in charge, and act that way.
- ❑ Strive to be respected, not necessarily liked.
- ❑ Always be available to your people.
- ❑ Cultivate patience. Know which things can be put off or ignored, and which can't.
- ❑ Pay attention to what your people are saying and to the back room talk.
- ❑ Never stop learning.
- ❑ Treat people consistently, fairly and firmly.
- ❑ Keep your eyes open, don't just look, but see.
- ❑ Make time to analyze problem areas, and compile facts before deciding.
- ❑ Those tough humbling experiences are valuable, treasure them.
- ❑ Give clear indications of what a good job is, give praise when it is achieved.
- ❑ Don't be afraid to acknowledge that you don't know.
- ❑ Quality can not be ordered, it is an attitude.
- ❑ There is a fine line between getting involved and getting in the way.
- ❑ Good supervisors surround themselves with good people and are not afraid of training replacements.
- ❑ Keep a positive attitude, keep company interests at heart.
- ❑ Set goals every day, review before leaving. Plan your days.
- ❑ Listen more, talk less. Be able to hear feedback you don't like.
- ❑ Solicit the views of the workers for improvements and problem areas.
- ❑ Use positive, one on one, techniques with workers.
- ❑ Follow your work plan.
- ❑ Keep your ass covered.

Supervisor personality assessment

What is your personality style?

We wanted to see what types of people were attracted to maintenance. We asked over 450 maintenance supervisors and managers to take a simple personality test. We used a self-administered version of a popular test called the Meyers Briggs Personality Inventory. This test (which might be used by your Human Resources Department), divides personalities into four dimensions. Each dimension can be divided into two traits. You score a series of questions and get a score in each trait. The higher the score, the more pronounced the trait.

The eight traits are:

I	introverted	**E**	extroverted
N	intuitive	**S**	sensor
T	thinker	**F**	feeler
P	perceiver	**J**	judger

Guidelines for the eight traits: Comment before each dimension is the broad category, percentage after category is percent in general population, the second percentage is the results of an analysis of the personalities of 450 maintenance supervisors:

I-E Shows how you choose to relate to the world, and where you focus your attention

I Introverted

25% in the general population and 51% in the maintenance population

Introverted or `I` types are likely to agree with the following questions:

I am reserved.

I like to puzzle out issues in my own mind before acting.

I prefer being alone or with one or two people I know well.

I communicate little of my inner thinking and feelings.

I tend to make decisions without consulting others.

I like quiet, thoughtful time alone.

Persons more introverted than extroverted tend to make decisions somewhat independently of constraints and prodding from the situation, culture, people, or things around them. They are quiet and diligent at working alone and are socially reserved. They may dislike being interrupted while working. They also tend to forget names and faces. Maintenance departments also have personalities. If the Introverted trait predominates in the departments, the style will be reserved, insulated, secretive, generate little or no consultation with people outside the department, and be misunderstood.

Possible strengths	Possible weaknesses
is independent	misunderstands the external
works alone	avoids others
is diligent	is secretive
works with ideas	is misunderstood by others
is careful of generalizations	needs quiet to work
is careful before acting	dislikes being interrupted

E Extroverted

75% in the general population and 49% in the maintenance population

Extroverted or `E` types are likely to agree with the following questions:

I enjoy discussing a new, unconsidered issue at length in a group.

I prefer activities and events which others join in.

I like talking freely for an extended period and thinking to myself at a later time.

I always try to make decisions after finding out what others think.

I like active, energetic time with people.

I enjoy meeting new people.

Extroverted persons are attuned to the culture, people, and things around them, endeavoring to make decisions consistent with demands and expectations. The extrovert is outgoing, socially free, interested in variety, and in working with people. The extrovert may become impatient with tasks that take a long time and does not mind being interrupted by people. An extrovert department (if there is one) will enjoy the interaction more than the maintenance work and will plan events where everyone can get together, have fun, and socialize.

Possible strengths	Possible weaknesses
understands the external world	has less independence
interacts with others	does not work without people
is open	needs change and variety
acts, does	is impulsive
is well understood	is impatient with routine

N-S How you see the world, how you acquire information

N Intuitive

25% in the general population and 43% in the maintenance population

Intuitive, or `N` types are likely to agree with the following questions:

It's fun to design plans and structures without necessarily carrying them out.

I like ideas.
I like the abstract or theoretical.
I enjoy thinking about possibilities.
I always try to think of new methods of doing tasks when confronted with them.
I am called imaginative or intuitive.

The intuitive person prefers possibilities, theories, the overall, invention, and the new, and becomes bored with nitty-gritty details, the concrete and actual, and facts unrelated to concepts. The intuitive person thinks and discusses in spontaneous leaps of intuition that may leave out or neglect details. Problem solving comes easily for this type of person, although there is a tendency to make errors of fact. To be successful, N type persons in maintenance will have to force themselves to deal with all the detail.

Possible strengths	Possible weaknesses
sees possibilities	is inattentive to detail, precision
sees gestalts (holistic view)	is inattentive to the actual and practical
imagines, intuits	is impatient with the tedious
works out new ideas	leaves things out in leaps of logic
works with the complicated	loses sight of the here and now
solves novel problems	jumps to conclusions

S Sensor
75% in the general population and 57% in the maintenance population

Sensor, or `S` types are likely to agree with the following questions:
I like to carry out carefully laid, detailed, plans with precision.
I want to know the factual details available.
I prefer the concrete or real.
I like to deal with actualities.
I like to use methods I know well, that are effective to get the job done.
I am called factual and accurate.

The sensing type of person prefers the concrete, real, factual, structured, tangible, here-and-now, and becomes impatient with theory, the abstract, and mistrusts intuition. The sensing person thinks in careful detail, remembering real facts, making few errors of fact, but possibly missing a conception of the overall. The sensor is great with details, and maintenance departments dominated by sensors will have policies and procedures for everything, in detail! Sensors also think everyone should follow the rules and become indignant when someone breaks them.

Sensor type personality:

Possible strengths	Possible weaknesses
attends to detail	does not see possibilities
is practical	loses the overall in the detail
has memory for facts	mistrusts intuition
works with tedious detai	does not work with the new
is patient	is frustrated with the complicate
is careful, systematic	prefers not to imagine the future

F-T which type of decision making is more comfortable

F Feeler

50% in the general population and 47% in the maintenance population

Feeler, or `F` types are likely to agree with the following questions:
I enjoy experiencing emotional situations, discussions, movies.
I like people who show feelings.
I have convictions.
I use common sense and conviction to make decisions.
I enjoy helping others explore their feelings.
I like being thought of as a feeling person.
I come to conclusions based on what I feel and believe about life and people.
I make decisions about people in organizations based on empathy, feelings, and understanding of their needs and values.

Feelers make judgments about life, people, occurrences, and things, based on empathy, warmth and personal values. As a consequence, feelers are more interested in people and feelings than in impersonal logic, analysis, and things, and in conciliation and harmony more than in being on top or achieving personal goals. The feeler gets along well with people.

Possible strengths	Possible weaknesses
considers feelings of others	is not guided by logic, is not objective
understands needs, values, feelings	is uncritical, overly accepting
is interested in conciliation	is less organized
persuades, arouses	bases justice on feelings

T Thinker

50% in the general population and 52% in the maintenance population

Thinker, or `T` types are likely to agree with the following questions:

I like using my ability to analyze situations.

I prefer verifiable conclusions.

I like logical people

I like being thought of as a thinking person.

I come to conclusions based on unemotional logic and careful step-by-step analysis.

I make decisions about people in organizations based on available data and systematic analysis of situations.

The thinker makes judgments about life, people, occurrences, and things based on logic, analysis, and evidence, avoiding the irrationality of making decisions based on feelings and values. As a result, the thinker is more interested in logic, analysis, and verifiable conclusions, than in empathy, values, and personal warmth. The thinker may step on the feelings and needs of others without realizing it, neglecting to take into consideration the values of others.

Possible strengths	Possible weaknesses
is logical, analytical	does not notice peoples' feelings
is objective	misunderstands other's values
is organized	is uninterested in conciliation
has critical ability	does not show feelings
is just	shows less mercy
stands firm	is uninterested in persuading

P-J how you handle time and what type of world do you prefer

P Perceiver

40% in the general population and 22% in the maintenance population

Perceiver, or `P` types are likely to agree with the following questions:
I like being free to do things on the spur of the moment.
I dislike using appointment books and notebooks but may use them
I start meetings when all are comfortable or ready.
I believe in planning as necessities arise, just before carrying out the plan.
I like change and I like keeping options open.
I review every possible angle for a long time before and after making a decision
I avoid setting deadlines.
I allow commitments to occur if others want to make them.

The perceiver is a gatherer, always wanting to know more before deciding, holding off decisions and judgments. As a consequence, the perceiver is open, flexible, adaptive, non-judgmental, able to see and appreciate all sides of an issue, always welcoming new perspectives and new information about issues. However, perceivers are also difficult to pin down and may be indecisive and noncommittal, becoming involved in so many tasks that do not reach closure that they might become frustrated at times. Even when they finish tasks, perceivers will tend to look back at them and wonder whether they are satisfactory or if they could have been done a better way. The perceiver wishes to roll with life rather than change it.

Possible strengths	Possible weaknesses
willing to compromise	is indecisive
sees all sides of issues	does not plan
is flexible, adaptable	has no order
remains open for change	does not control circumstances
decides based on all data	is easily distracted from tasks
is not judgmental	does not finish projects

J Judger

60% in the general population and 78% in the maintenance population

Judgers, or `J` types are likely to agree with the following questions:
I like setting a schedule and sticking to it.
I get the information I need, consider it for a while, and then make a fairly quick, firm decision.
I plan ahead based on projections.
I keep appointments and notes about commitments in notebooks or appointment books as much as possible.
I prefer knowing well in advance what I am expected to do.
I tend to start meetings at the prearranged time.

The judger is decisive, firm, and sure, setting goals and sticking to them. The judger wants to close books, make decisions, and get on to the next project. When a project does not yet have closure, judgers want to leave it behind and go on to new tasks and not look back.

Possible strengths	Possible weaknesses
decides	is unyielding, stubborn
plans	is inflexible, unadaptable
orders	decides with insufficient data
controls	is judgmental
makes quick decisions	is controlled by task or plans
remains with a task	wishes not to interrupt work

Building personality from combinations of traits

When the traits are combined into one personality, various aspects strengthen and modify each other. For example some two letter combinations:

NT 12% in the general population and 19% in the maintenance population: There are four of these including ENTJ, INTJ, ENTP, and INTP. These people tend toward abstract, analytical thinking, and are attracted to new ideas and applying logical processes. They tend to be visual, visionary, competent, bookish, and sometimes arrogant

NF 12% in the general population and 24% in the maintenance population: As with any of the couplets, there are four versions: ENFJ, INFJ, ENFP, INFP. These people like relationships and are attracted to warmth and possibilities. They like new projects and new people, are optimistic, like experiences, and would be excellent journalists or coaches

SJ 38% in the general population and 44% in the maintenance population: As with any of the couplets, there are four versions: ESFJ, ISFJ, ESTJ, ISTJ. These people are practical, organized, fair, consistent, and feel comfortable operating in environments with rules.

SP 38% in the general population and 12% in the maintenance population: The final couplets also have four versions: ESFP, ISFP, ESTP, ISTP. These people are spontaneous, flexible, make good fire fighters (real fires), they get immersed in what they do, and love crafts.

Descriptions of the 16 personality combinations and their frequency in the maintenance population:

ISTJ 14.5%

Maintenance persons that are well organized, serious, and quiet. Respect for and facility with facts. On the surface they are calm in crisis but might have vivid reactions underneath. They take responsibility, and will make up their own minds about a situation, working slowly and methodically, regardless of protests or distractions. They tend to stabilize projects and work groups. They are at risk of thinking everyone is like them and overriding less forceful people. They might also minimize imagination and intuition.

ESTJ 13.8%

Very organized and like to set goals. Practical maintenance persons who are realistic and matter of fact. More interested in the here and now. Naturally good mechanics, they like to organize and run things using logical processes. Make natural maintenance supervisors when they can consider feelings and values of others, though they may decide too quickly.

INFJ 10.2%

These supervisors will succeed through perseverance and originality. Strong idea people driven by inspirations. They need to be careful to avoid being smothered by routine aspects of the job. They put their best efforts into work. Quietly forceful, conscientious, concerned about others. They are respected for their firm principles, and their desire to do what is needed or wanted. Their inspirations must be tempered with development of their thought process.

ENTJ 10%

These supervisors are hearty, frank, decisive, and well-informed leaders. They enjoy long-range planning and thinking ahead, and are good at logical, intelligent, reasonable, speaking. Their main interest is in seeing possibilities beyond the present, though they may be over-optimistic for the situation. They also need someone around with common sense because they may decide too quickly and sometimes ignore others' values and feelings.

ENFJ 8.5%

These supervisors are sociable, popular, and responsive to both praise and criticism. He or she is persevering, orderly, and conscientious, and interested in possibilities and harmony. They also are concerned about what others think, want, and feel, and can present proposals or facilitate groups with ease and tact. They might tend to jump to conclusions, and have many rules specifying shoulds and should nots.

ESFJ 8.1%

This group are warm-hearted, outgoing maintenance supervisors, who are born cooperators. They have little interest in technical areas, but will create harmony and find value in other people's opinions. Their personalities are practical, orderly, and down to earth. They work best in a warm encouraging, supportive environment, but have many rules and assumptions about situations.

ISFJ 7.9%

These supervisors are quiet, friendly, responsive and dependable, and will work devotedly to meet obligations, lending a stable influence to any team. They care about the people on their crews, and are thorough, and painstakingly accurate. They need time to master technical areas, and are at risk of becoming withdrawn. They do not trust imagination or intuition

INTJ 4.2%

These maintenance supervisors have original minds and great ideas, with strong intuition. They have the power to organize and carry out jobs with or without help, and will drive others as strongly as themselves. They are considered to be stubborn, skeptical, and independent, but they are single-minded and should seek and need the input of others to balance them. They need to develop their thinking so that they can evaluate their own inspirations.

ISFP 4%

Quiet, friendly, sensitive, modest, kind, types of maintenance supervisors. Might not show warmth until they know you well. They look at the world through their own deeply held values and don't like disagreements. They are craft-oriented, loyal followers, and do not usually choose to lead. They like a slow pace, so they may not measure up to their inner ideals, causing them to feel inadequate.

ESFP 3.6%

These supervisors are outgoing, easy-going, accepting, and friendly, and make very flexible problem- solvers, not bound by current rules and procedures. They are actively curious about people, objects, food, and anything sensory. They know what going on through the grapevine and participates, and they remember facts better than theory. They learn by doing, and have good common sense, though they may sometimes be too easy on discipline. Their love of a good time might put them at risk.

INTP 3.6%

These quiet maintenance supervisors are extremely logical, abstract, and interested in ideas, having sharply- defined interests. They are more interested in ideas than in their practical application. They have little need for social small talk, get-togethers, or anything else that is not in their area of interest, and they sometimes overlook other people's values and feelings.

INFP 3.4%

These people are enthusiastic maintenance supervisors, little concerned with surroundings or trappings, and deeply driven by inner convictions that are usually difficult to express. They enjoy learning, ideas, languages, and independent projects. They can be very absorbed in current activity, and are sometimes excellent writers. They usually are not usually talkative until they know you well, and they often feel they don't measure up to their own inner standards.

ISTP 3.2%

These supervisors are cool, quiet, and reserved people who analyze everything with detachment, and prefer to organize ideas and facts rather than people or situations. They are interested in impersonal principles and why mechanical things work, and may have great capacity to understand the facts of a situation. They are engineering oriented, but will not overexert themselves so they may overlook other people's needs and values. They also may not follow through well.

ENTP 1.7%

These supervisors are quick, ingenious, energetic, and good at many things. Their thinking helps temper their intuition. Although they may neglect routine assignments, they are resourceful at new, unique, and novel problems. They make stimulating companions and may argue either side of an issue for fun. They must constantly feel challenged, but need to learn to follow through. Without development of their judgment they may waste their energy on ill-chosen projects.

ENFP 1.5%

These people make warm, enthusiastic, maintenance supervisors, having can-do attitudes and able to see possibilities everywhere. They are quick to help and to suggest solutions. With their great insight, they make skillful people-handlers. Becasuse they hate routine, they will improvise rather than plan in advance. They may leave projects after the core problems are solved because they become bored and may leave projects uncompleted.

ESTP 1.2%

These types make no-worry, no-hurry supervisors. They operate from concrete reality - what they can see, hear, touch, taste, or smell. They enjoy sensory pleasures and are not bound by current rules and procedures to find the solutions to problems. They like mechanical things and movement. Blunt, and occasionally insensitive, they are natural craftspersons when they can include people. Even after they become supervisors, they keep a mechanical hobby at home for the joy of it. Their love of a good time might be a risk.

Hints on applying the above inventory to individual situations

1. People who have the same strengths in the dimensions seem to click together; and they often arrive at decisions more quickly. They seem to operate on the same wave-length. They work from similar assumptions.

The problem with people who have the same strengths in the dimensions is blind spots. Their decisions may suffer because they have similar weaknesses.

2. Different traits have different strengths: the intuitive is best at seeing the future, practical realism from the sensor, analysis from the thinker and the skillful handling of people from the feeler. Groups with a preponderance of members having similar strengths should seek out and listen to other types when making decisions. This openness is essential for major decisions.

3. People with different strengths may not see eye to eye on many things. The more the group differs, the more likely misunderstandings and conflicts will occur. The advantage of groups with different strengths is that better decisions may result from the differing points of view.

4. People might be sensitive about criticism in their areas of weakness. These people might avoid using their weaker sides and conflict may arise when they must use those dimensions, or when others point out deficiencies.

5. A person's topology cannot be changed to its opposite. Each person can realize where their weaknesses lie and work to strengthen them.

6. People's values, beliefs, decisions, and actions will be profoundly influenced by all four of their stronger dimensions.

7. If you want to look into this test as a team building tool, contact your Human Resources dept and ask for the Meyers-Briggs Personality Inventory. Human Resources should have copies and information on its use.

Communications and Delegation for Maintenance Leadership

Why study communications? We study these techniques because:

- ❑ Communications supports management objectives.
- ❑ Good communications is a skill that can be learned.
- ❑ You will feel better about yourself if you are a better communicator.
- ❑ You will be a more effective delegator and a more effective manager.
- ❑ Your family life will benefit.
- ❑ It's fun.

Communication does not come free. Of course, some people are born with more innate skills but like skiing, welding, or any human activity, everyone can learn to improve. Some of the things that communications requires are:

- ❑ Self discipline to pay attention to all the details of the interchange.
- ❑ Knowledge and skills of communications techniques.
- ❑ Awareness of what is going on with the other person.
- ❑ Reason to communicate.
- ❑ Intention to communicate by all parties.

Communications scientists tell us a surprising fact: that 70% of the message in a communication comes across through non-verbal channels such as voice tones, voice volume, body positioning, facial expressions, animation, air space, eye contact, fluency, physical barriers, hand gestures, skin coloration, breathing patterns, and others.

On the shop floor there is a critical communication issue:

Take note of cross-cultural/ethnic/sex/class communication barriers. When dealing with people from cultures different from yours, be aware of your own assumptions about them. Try (it's difficult) to see what stereotypes you have about them. One trap is golden rule thinking, which says I'll do unto them what I would like to be done unto me. The problem is that different cultures think and act differently enough that you might be deeply insulting or at the least, misinterpreted. Always be on the alert for unintended meanings, unfamiliar gestures, and different customs.

Maintenance is a multi-ethnic, multi-cultural, multi-class field. It is almost inevitable that members of a maintenance workgroup will be diverse. There are many excellent books that highlight the pitfalls of assuming that your communication style is appreciated, or even understood, by your co-workers. A deep discussion is beyond the scope of this text, but a serious study of these topics is recommended for maintenance leadership.

Communications allows you to turn problems into Opportunities

Your customers enter the maintenance sphere of influence via a work order, service request, or other, less formal means. A good system will insure that the vital communication of what needs to be done is complete and accurate. It takes discipline (write everything down and ask all relevant questions), knowledge, and skills to translate what the customer is asking for into what is needed from maintenance. The person interacting with the customer needs some sensitivity to the unspoken condition of the other person. Note that in the chapter on work orders we introduced the concept of a class to teach how to fill out work requests. That type of class will eliminate many communications problems.

Look at your work request system as a communication opportunity. Any time a mechanic does the wrong job, at the wrong place, to the wrong unit, is an opportunity to look at the effectiveness of your work initiation system. Any time there is a wild goose chase, when a mechanic is sent out with the wrong tools, or the wrong materials, or the wrong expectations is an opportunity to review the workings of the internal communication effort.

When a breakdown occurs in communication, most supervision effort has traditionally been oriented toward discovering who the person was who "screwed up." Most of these breakdowns occur because some part of the communication system doesn't work well. If you investigate the defect in the system that allows mistakes to happen, you will have the opportunity to improve the whole system.

For Fun take an Inventory of your current communication skills

Scoring:	5	4	3	2	1	0
I am:	Very Strong	Strong	OK	OK	Weak	Very Weak

Answer the following questions as honestly as you can:

1. I am relaxed when I am communicating: _____
2. I look people in the eye while talking: _____
3. I lead the discussion: _____
4. I encourage the other person to speak rather than giving my opinions: _____
5. I seek (rather than give) information: _____
6. I tend to summarize discussions rather than evaluate who was at fault: _____
7. I ask open questions: _____
8. I let communication happen anywhere, rather than only in my office at my convenience: _____
9. I talk minimally to encourage the other person to talk: _____
10. I make it easy for the person to talk: _____
11. I reflect back on how the person must have felt: _____
12. I can catch subtle clues in a person's speech, manner, or gestures _____
13. I can stay with someone in a conversation: _____
14. I have good timing in communications: _____
15. I will clarify rather than confuse a person: _____
16. I keep communications on target: _____
17. I know what my people are talking about: _____
18. I accept all my people just the way they are: _____
19. I am aware of myself when communicating: _____
20. I express my feelings when communicating: _____
21. I stay on the topic at hand: _____
22. I am aware of information on the non-verbal channels of communication: _____
23. I can tolerate and use silence in communications _____
24. I stick with a problem and help define it: _____
25. I am effective in working with people's problems _____
26. I can operate at deep levels of analysis _____
27. I understand people _____
28. I can restate what people say to demonstrate my understanding of what is being communicated _____

29. I catch the essence of and can summarize a communication: _____
30. I build a strong bridge of rapport: _____
31. I provide a supportive environment for my people: _____
32. I trust my people and they trust me: _____

TOTAL SCORE: _____

A perfect 100% communicator — *160 points*

This person should write a book about communications. A perfect communicator possesses great disciplinary skill and awareness. Their star will rise (or has risen) in most organizations.

An `A` level communicator — *140 points or above*

Many top executives are superior communicators at this level. They got to the top through people who would work for them and with them. Their ideas and visions were properly transmitted. More importantly, their people felt that they had been listened to when they spoke to these high level communicators.

A `B` level communicator — *125 points to 139 points*

This good score shows some need to improve but that great territory has been covered. A sharpening of skills to the elite level will take work and time. A `B` level communicator will rise in their organization.

A `C` level communicator — *105 points to 124 points*

This is an average communicator. Much work has been done but there still is much to be done. Chose an area and concentrate on it for a month or two. Then choose a new area and concentrate on that. After a year, see if your scores don't improve substantially. With the improvement in skills may come more responsibility as people see the results of the effort.

A `D` level communicator — *90 points to 104 points*

Face the fact, you have work to do. You could be a great engineer, electrician (or whatever), but your message will not get across to your people. They won't understand you and won't feel understood by you. Your impact on the organization is minimized by your inability to get your vision and your point across. Many people in this situation are more comfortable with machines, drawings, and the technical aspects of the job. Follow advice for C and expect significant gains in a short time.

An `F` level communicator *89 points or less*

Maybe you just started needing to communicate. You may need coaching in reading, test taking, or English. This may explain the low score. There is nowhere to go but up. Keep your head up and follow the advice given to `D.`

Delegate to Thrive

"Supervisors are not paid for what they can do but for what they can control."

In Webster's New World College Dictionary delegation is defined as: "to entrust (authority, power) to a person acting as one's agent or representative." The first ground rule of delegation is to entrust both the authority (power) and the responsibility.

The maintenance leader's job is to work through other people. One of the most difficult transitions is from worker (being paid for how well you work) to supervisor (being paid for how well you work through others). One of your main jobs is the development of the people that work for you. Delegation will help develop talent within your work group. The effect of some responsibility on people is amazing.

Many leaders resist delegating work to their subordinates. Often the reason is that the leader is sure that they can't handle it, or that he or she can do it better. Other fears might be:

- ❑ The subordinate might mess up the job and cause a bad reflection on the supervisor
- ❑ Loss of control
- ❑ Feeling threatened by having to train a replacement
- ❑ Looking bad because other people are doing the `actual` work
- ❑ Supervisor's often like their jobs because they like to feel needed (always being the center of attention).
- ❑ Delegation might reduce their position.

On the other side of the coin, the subordinates might resist the assignment because they feel that they are being set up, they are already too busy, and they have learned that it's safer to rely on the supervisor.

Do's and Don'ts of What to Delegate

DO	DON'T
1. Routine Tasks 2. Time-Consuming Jobs 3. Skill Improvement Tasks	1. Personnel Tasks 2. Job Assignments 3. Disciplinary Actions

Letting people make mistakes

Someone said that experience is gathered as a result of bad judgment. Whether you agree with this thought or not, people have to be allowed to make mistakes, fail, and have bad judgment. Employees who only rely on the good judgment of an experienced supervisor or leader (instead of themselves) stunt their growth opportunities.

Through delegation, create a safe opportunity for mistakes. You become the sidelines coach. Allow enough emotional space for your people to make their own decisions, which will help them to grow. Keep yourself from interfering (we already know you can do it better). As long as they are not in danger (to themselves, you, large batches of product, etc.), let them learn, it will make them better employees. Consider a relief supervisor slot as a training ground for future supervisors when you vacation (or go to seminars).

The One Minute Manager

The One Minute Manager is still one of the most popular self improvement books for managers and supervisors even though it was published in 1983. The rules apply directly to maintenance leadership. We strongly encourage all maintenance leadership people to purchase and read this excellent and very short book. It contains three concepts which are simple and powerful.

To maximize the effect of The One Minute Manager concepts, discuss the goals and concepts with your work group first. Encourage people to read the book (circulate some copies). Discuss the concepts so that everyone knows what you are trying to accomplish. The authors of the book (particularly Ken Blanchard), have created an extensive library of books, tapes, and seminars to teach these concepts and their more recent material.

1. One minute goal setting: In delegation, most of the problems stem from the person not really knowing the what, when, how of what you want. Make it clear what they are to do. Have them write the goals (or scope of work) out on a single sheet of paper. The total statement should be less than 250 words. In

the words of the author "feedback is the breakfast of champions." The goals should be written in the first person (using I), and in the present tense. The people should read their goals every morning or whenever they start to work on the project (should take less than 1 minute). In simpler delegations the goals might correspond to a scope of work.

2. One minute praising: Catch people doing something right! Look for approximately right. Remember, feedback is necessary for personal continuous improvement. This request is deeper than it seems on the surface, and it goes against the training and culture of maintenance. Maintenance people have cultivated their ability to see what's wrong. They can walk past 10 presses and hear the one that needs grease. The ability to see subtle deviations may save a machine or even sometimes a life. This ability also makes it difficult to see what's right with people's work. The one minute praising is essential for maintenance professionals who want to improve the impact of their leadership.

Steps:
1. Tell the person what they did right. Be very specific.
2. Tell the person how that makes you feel or the impact of his act.
3. Do it now.

3. One minute reprimand: The authors use the word reprimand, which sounds like a serious discipline action that might involve the union or higher management. Reprimand is actually a special type of conversation you would have with a subordinate, with the intention of putting them back on course. It does not involve discipline at this point. The reprimand is also an opportunity to express your anger and frustration before it can build up and become destructive. The person gets any guilt relieved and knows what is expected.

Idea: we are not our behavior. Before you reprimand, be sure you have the facts. Give the reprimand in a private location.

Steps:

1. Be specific, tell people exactly what their behavior was that made you angry.

2. Tell them how that behavior made you feel.

3. Allow a pause, make sure the person understands that their behavior was the issue, not them as a person.

4. Do it at the time of the occurrence, not later.

Three thoughts on The One Minute Manager

1. Do you believe that you have to be great at these precepts for them to work? The fact is that they will work if you use them. You will improve over time.

2. You will increase productivity because you will release people's intelligence and improve motivation. The better your people look, the better you look.

3. Good productivity is a journey, not a destination.

Time Management in the Maintenance Pressure Cooker

Maintenance can be a pressure cooker. Emergencies, short staffing, vendor problems, high customer expectations, all contribute to stress on the job. The stressors are in the environment. Stress is your body's reaction to the stressors. Although the stressors cannot be removed, changing your attitude and reducing your stress level might be possible. A person who feels mastery over his/her environment transforms the `bad` stress into good excitement.

Time management helps people feel mastery. The goal of time management is to make you feel that you can master anything that comes at you.

Effective Time management is a four-pronged approach to gaining control of your day while multiplying your output, and it is accomplished by the following actions:

1. Evaluate and shift how you think about time
2. Four projects that show you new ways to work and think about your time
3. Seven daily habits of successful supervisors.
4. Twenty strategies for dealing with time killers.

The Nature of Time- how do you spend it?

Time is the only truly non-renewable resource. Unlike energy, time cannot be saved or created. Time is also the only resource of which everyone has the same amount. A leader of a multi-billion dollar organization has exactly the same amount of seconds and hours in the year (31,557,600 and 8766 respectively) as you do. In the amount of time we are all truly equal.

TIME MANAGEMENT = SELF DISCIPLINE

Time management is a life skill. A life skill is a skill that managers should hone and sharpen throughout their entire life. However high you rise in your

Preliminary Self Test of TIME MANAGEMENT

1. Are you satisfied with the way you spend your time?	yes	no
2. Are you satisfied with the number of hours you work?	yes	no
3. Do you enjoy your work?	yes	no
4. Do you feel good about the quality of your work?	yes	no
5. Do you effectively cope with the stress/pressure of your job?	yes	no
6. Do you have enough free time to pursue hobbies, community activities?	yes	no
7. Do you have time to enjoy your family?	yes	no
8. Are you satisfied with the results you achieve?	yes	no
9. Are you satisfied with your bosses' concern for time management?	yes	no
10. Are you satisfied with your subordinates'concern for time management?	yes	no
11. Does your use of time reflect your goals ?	yes	no
12. Are you satisfied with the quality of time spent with your subordinates?	yes	no
13. Are you an effective delegator?	yes	no
14. Do you react to changes in a constructive manner?	yes	no
15. Are you in control of your telephone?	yes	no
16. Do you take advantage of time windfalls?	yes	no
17. Are you well organized?	yes	no
18. Do you have a well designed plan of organization for your desk?	yes	no
19. Do you take time to exercise?	yes	no
20. Do you occasionally do nothing?	yes	no

Score 1 point for each yes, 0 points for each no

Score range

18-20: Could (and probably should) teach time management courses in your spare time.
15-17: Good time manager
12-14: Above average in time management
9-11: Pay close attention to this chapter!
0-8: Tough row to hoe, is your head still above water?

organization, and wherever your life takes you, time management will make you more effective. Good time managers also have time (and sometimes more importantly, energy) for the other fun things in life like family, hobbies, and contributing to their community.

Other strategies for finding out how you spend your time

One of the most effective methods of tracking activity is work sampling. The key to work sampling is to randomly note your own activity throughout the day. Make a note of the activity at each instant of the random time.

Reality of being a maintenance supervisor: The reality of the field of maintenance is that frequently we are not in control of our time. Please note that when we are not in control of most of our time it is doubly important to control what little time is left.

Do you know exactly how you spend your day? Using the form on next page, take a minute to write down the percentage of time you spend at each activity in an average week. Please try to note what you actually do, rather than what you'd like to be doing (or what you'd like other's to think you're doing).

As you consider the percentages, ask yourself if you are the kind of supervisor you'd like to be. If you're not, ask what that supervisor would be spending time on. In the second part of the exercise, fill in these `ideal` times using the right column.

Use a Log Sheet from a day planner or the computer:

When you return to work, keep a log of what you really do for a week. Time management experts agree the place to start managing your time is finding out what you actually spend your time doing. Copy the log sheets for as many days as you plan to log. The sheets should be filled out as the day progresses (NOT AT THE END OF THE DAY).

Daily Activities exercise

Activity	Your Percent	Best Manager
Budgeting	______	______
Circulating (moving through your domain, insuring work is done and done correctly)	______	______
Dealing with users/customers	______	______
Engineering jobs	______	______
EPA/Hazardous waste activities	______	______
Failure analysis, other deep thought	______	______
Giving job assignments	______	______
Inspecting work	______	______
Looking for information, drawings etc.	______	______
Meetings	______	______
Other regulatory activities	______	______
Quality circles, other group efforts	______	______
Purchasing/parts related activities	______	______
Paperwork (all other)	______	______
Personal activities (all non-company activities on company time)	______	______
Reading junk mail	______	______
Scheduling, planning	______	______
Seeing salespeople	______	______
Teaching, training	______	______
Wrench turning' time:(all physical work activities)	______	______
Other ________________________________	______	______
TOTAL	**100%**	**100%**

Figure 61

Time Management Projects

Project #1 Clean your desk, organize your office.

A. Set aside a length of time when you won't be interrupted.

B. Get an entire box of manila folders. Put them in order with the tabs alternating left-center-right-left-center-right.

C. Get a blank piece of paper (full sized) for your master To-Do list.

D. Put all your papers in a pile. Include the little half- and quarter-sheets, envelopes, napkins, and everything else you were going to get to.

E. Go through every piece of paper, and be ruthless about throwing out as much as possible. Separate into a new pile, the papers you need to keep. If there is an action that needs to be taken, save the paper in this new pile.

Now rest for a few minutes. All your old piles have now been consolidated into a new smaller pile. In fact, maybe 1/2 of your old stuff is now in your trash can.

F. Back to the grindstone. Start with the top paper and ask yourself some simple questions:

Is there any work that must be done? Add that assignment to your master to-do list.
Ask, should I keep the piece of paper? Remember you recorded the assignment. If it's a keeper, and no file already exists, prepare a manila folder.
Keep this up until you reach the bottom of the pile. Do not do the work now unless it can be handled within 1-2 minutes.

G. Remove all office supplies such as tape, staplers, and paper clip collections from the surface of your desk. Attack any surface of your office that accumulates papers for more than a very short time (the paper turns over in 1 shift).

H. Apply the same standard to your files and drawers. Review the information and determine if you need it, and if there is any action that you need to take. Clean your drawers of the debris that accumulates over the years.

I. You now have a clean desk. You also have a completed to-do list that could be up to several pages long. You will find that a clean desk encourages efficient work habits. You put your work on your desk, complete it to the level possible at that time, put it back into its proper files, and have a clean desk once again.

Project #2, Supercharge your time.

You have three major tasks:

1. Get your immediate job done.
2. Have time for the large important jobs and analyses.

3. Educate yourself to improve your value to your company (Stephen Covey calls this sharpening the saw)

A. Look at areas where you could do two things at once without sacrificing the quality of either. An example would be the twice daily commute. A one-hour commute could translate into over 200 hours a year of learning time through use of cassettes. Remember that 200 hours is equivalent to three college level courses (homework time included). Ideas: waiting for meetings, waiting for airplanes, driving, exercising, taking a bath, mowing the lawn, etc.

B. Carry a micro voice recorder to record ideas, memos, letters, and instructions. This idea is excellent if you have staff support to transcribe your tapes. Of course, if you record in digital it might be possible to process the file by a voice recognition system, eliminating the need to transcribe the notes. The transcription should use a separate page for ideas on each project, to be filed in the corresponding project file. Your brain is your most powerful tool.

C. Delegate some of your tasks to your crew. A well trained crew will multiply your effectiveness.

D. Learn to type faster and use a computer better. Once you get over the hump, you will type as fast or faster than you write, with less fatigue and greater accuracy.

E. Use fax machines (or scanners and E-mail) to transmit ideas (particularly drawings) to remote sites, vendors, and users. Faxes are faster than mail and more accurate than any verbal descriptions.

Time yourself reading the next paragraph

Start timing: G. Learn to speed read. If you think that might be a waste of time, try this: Time your reading speed. Read this whole paragraph to the bottom of the page fairly closely at your normal reading speed. Note the elapsed time. Divide the total word count (106) by the number of seconds and multiply by 60. A good fast pace is 400 words per minute with good comprehension. Speed readers top out at 1500 to 2000 words per minute. Many maintenance supervisors read at 100 or less words per minute. How much more could you read if you could read at 1000 wpm? Check below. **Stop Time...**

(Note: Paragraph G has 107 words. A speed reader would have completed it in 6-10 seconds.)

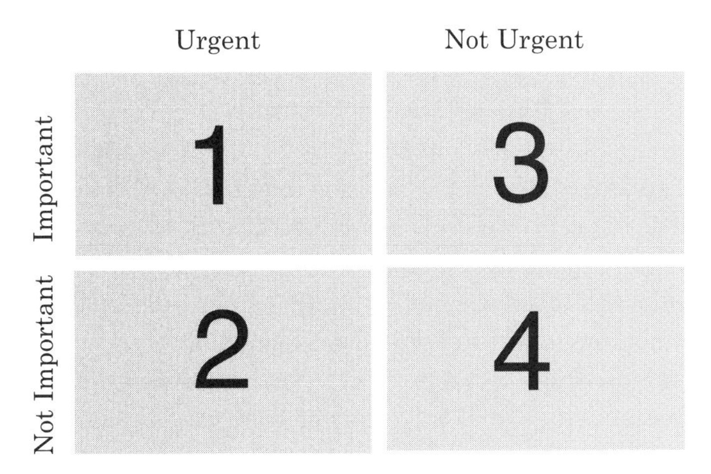

Figure 62 The four types of tasks

#	Master To-Do List	Topic	Date On	20%/80%

Project #3 The 80/20 Rule

A nineteenth century Italian economist, Vilfredo Pareto, studied the distribution of wealth in Italy. He found that very few families controlled most of the wealth. His research morphed into an informal rule called the Pareto Principle or the 80/20 rule. This rule can be stated that for most management situations, 80% of the action comes from only 20% of the actors. We find that 80% of our employee problems come from 20% of our employees, 80% of our emergencies come from less than 20% of the equipment.

Thus, 80% of our results will flow from 20% of our activities. If we identify these few critical activities, and increase our time commitment to them, we can double or triple our results each day. The cream of the crop is your bottom line activities. These are defined as activities that you will get done no matter what.

Your exercise is to go through your master to-do list and examine each item. Divide the items into a 20% important and highly-leveraged activity and the 80% low-leverage activity.

The 20% activities should be scheduled first during your high-energy intervals throughout the day. The top three of these activities become your bottom lines. The bottom line is `if you do nothing else today, you will do your bottom lines`.

*******Beware of the tyranny of the URGENT.**

Remember "You are paid for what you complete not what you start.

One of the greatest thinkers and consultants of this era is Dr. Stephen R. Covey. In his excellent books, Covey describes a new way to sort out events that exactly corresponds to good maintenance practices. His model reproduced on the previous page describes four kinds of incoming tasks divided into two categories:

1: Most maintenance managers are already excellent at category number one, which is called fire fighting. I've never yet met a maintenance professional who would say they were not great fire fighters. The bad fire fighters probably get winnowed out and go into an allied profession that doesn't have as many crises. You need to be an effective fire fighter to keep your job.

2: This Unimportant-Urgent category is the death of the maintenance professional. Urgency is implicit in most maintenance requests from users. The phone rings and everything sounds important (particularly to the caller). It is tough to separate the important from the unimportant. No one was ever fired from a maintenance job because they responded too quickly to the urgent-unimportant jobs coming into the maintenance center.

3: World class maintenance requires significant attention to events that are important but not urgent. This category includes the whole PM system, your training efforts, and your operator involvement programs. Usually, no one is yelling for you to interrupt production to perform PM's. PM's are essential for reduced breakdowns but it is not pressing for them to be completed.

4: This Unimportant-Not Urgent is an easy category. You can see the requests stacked up on maintenance professional's desks, on top of file cabinets, in closets. These are the requests that you've been meaning to look over but never seem to get time, and they are the fun, brainless stuff, like junk mail. The category is not really important and definitely not urgent.

The goal of time management for maintenance professionals is to shift time from category 2 and 4 into category 3. Category 3 reduces the category 1 problems for the future.

Project #4 Energy: Establish your energy level by the time of day.

In this important project, identify the times of the day when you have superior concentration and energy. At the same time, identify times when your concentration is lacking.

Schedule the Important-Not Urgent tasks at the high energy times. Schedule the 20% high return on investment activities to be performed at these times

Seven daily habits:

These habits will help you focus your time and energy so that the important things get done. For most people, identification of the high return on investment activities is difficult. The seven habits institutionalize processes to identify the important things each day.

These seven daily habits help your time management in several ways. These habits help use your whole mind through utilization of ideas or aha's. The habits ensure balance, so that you spend time on others, and yourself, in addition to work. They also give you points for exercise, diet, and rest.

1. The first activities help you set up your day. You approach each day with resolve and calm. In the whirlwind of everyday life, we forget some things. Without a focus we become slaves to the urgent, and never see the important things that float by.

Spend some quiet time and do the following three activities. Some people schedule these activities the night before. Others work on them early in the morning.

a. Purpose: Write a short description of the coherent purpose of your work situation (keep this private). Read and review this purpose every day. The purpose should cover what you are moving toward. Some people like to look at this declaration as their personal mission statement. The statement could include information about your purpose at work and at home, information about what you want for your family, the importance of spirituality in your life, and how you expect to contribute to the world.

b. Planning: Spend a few minutes each morning (or the night before) planning your day. Consider your purpose statement. Review your bottom lines and see what elements can be incorporated now. Under the best conditions, your plan may go down the drain. Having a plan increases the probability that there will be a positive outcome.

c. Preview: Visualize your day ahead of time being the way you would like it to be. Practice in your mind's eye. Many great sports figures practice in their mind before they swing the club, the bat, or shoot the ball. They see success. We need to look at our work plan and see success in the elements. We see in our powerful mind the successful meeting with the boss or a solid coaching session with one of our people.

2. Feared: Your first task every day is to schedule and do the thing in your plan that you fear or resist the most. Fear stops people from being outstanding contributors to their organizations, communities, and families. If each day we face one of our little fears, we would move forward quickly with our agendas. . It is said that anything is achievable if fear can be felt and then overcome.

3. Self, Other, Work: Schedule and complete the three most important tasks, called bottom lines, to support your purpose in the following areas: Work, Others, Self. These daily bottom lines should be done each day, even if nothing else goes right. By doing the bottom lines even on the `bad' days, you move closer to completing the important but not urgent activities.

4. Nutrition, Rest, and Exercise: High efficiency depends on a healthy body. Be sure to plan for adequate rest, good nutrition, and daily exercise.

5. Idea: Write down any ideas that come to you about your situation or life in general (work or home). These insights are called ahas. . Your brain is more powerful than the biggest computer made and is always switched on. Many insights are outcomes of thought processes that you are not aware of. Noting these insights enriches your life and uses the brain more effectively.

6. Be a teacher or mentor to another person. Every day, one element of your mission is to return something to the people around you. Schedule yourself to be in a position to contribute, mentor, or coach another person, every day.

DATE

Day's Theme

Purpose		Nutrition	
Planning		Rest	
Preview		Exercise	
Feared		Idea	
Self		Mentor	
Others		Review	
Work		TOTAL	

Today's HOT List 20% activity		
	Feared	
	Self	
	Other	
	Work	
	Exercise	

Today's 80% activities not-hot		

Ideas/Aha's

Schedule
6am
7am
8am
9am
10am
11am
12noon
1pm
2pm
3pm
4pm
5pm
6pm
7pm
Evening

7. Review your day to see where you did well and where you would like to make changes. This study is like reviewing films of last night's game. The coach looks at the film with the players with the intention of improving play. A good coach will say what is necessary to help the players improve, including congratulations, yelling, suggestions, and anything else that would modify behavior.

The balanced life gives long term satisfaction

Ideas for bottom line activities. To balance out an active and engaging work life, these other bottom lines have to be scheduled.

FAMILY: One of your bottom lines is in serving your family in some way. Even if you don't have children, the rest of your family is important for balance. Balance is essential for long term stability, health, and some level of happiness. An example would be to take your kids on a hike once a month.

BEING OR BRAIN: Improve yourself. An important bottom line is to be investing in yourself. This investment could be in the form of reading a book, taking a course, trying something new. An example would be to read one book this quarter on maintenance management. The being also includes your spiritual life. It could be to pray, attend services, or other religious activity.

BODY: Your body is your vehicle to anything that you want to get done. Health cannot be guaranteed. You could exercise every day and eat right, and still have health problems. Healthy activities can contribute to your quality of life, and can help keep you healthy. Items on this bottom line might include walking for at least 30 minutes, five times a week.

TIME OR DELEGATION: There are many specific ways in which we could improve our delivery of service. We need to invest some of our time increasing our time effectiveness. This effort could include training on a new computer program that will make your time more effective. The bottom line could also be to delegate more activity. To delegate you might have to train a subordinate or seek out a new vendor.

PEOPLE: Successful people rarely are total loners and they usually have a group of people whom they trust, that they can go to for advice. The bottom line fixes your attention on the people aspect of success.

MONEY OR WORK: You thought this was all that there was to life. Your company might agree.

Taming 20 time killers

1. Take control of some of the small parts of the high energy time of your day. Prevent interruptions during these times by rerouting calls to a crew member, clerk, or other supervisor. Use these times for bottom lines and high return on investment activity. Guard your high energy times.

2. Some meetings are energy killers. Never schedule meetings during

your high energy times. The exception is the 15 minute `hit the ground running` meetings. The best time for meetings that you control is around lunch time and at the end of the day.

3. Train yourself to be able to put things away and throw things away. You may walk around a pile of magazines for months before realizing they should be in the circular file. Look around your work space and once a month, throw some junk away. Do it until only current or important stuff is left. This activity is a good filler when you are low in energy or at loose ends, with a few minutes to spare.

4. Get used to the fact that you don't have all the answers. When something stumps you, restate the problem, spend time trying to isolate the core of the problem. Seek out people with knowledge, both inside and outside your firm. Focus on gathering information rather than looking for answers.

5. Attack your overweight bulging Rolodex, E-mail directory, Outlook address books. Business would screech to a halt without these organizers. The goal is to have a slim address book that is quick to use. Limit it to your current most-often-used numbers. Develop the habit of writing or stamping the date on every record. Make sure every number and address you use regularly is in your system. If you have a speed dialer, set it up now.

6. Put throw-out (T.O.) dates on all files. These T.O. dates will keep your files clean. Every six months, review your files and throw out the old ones.

7. Try this if you get magazines that come only to you: when a trade journal comes in: read the table of contents. Rip out the articles you want to read and put them in a reading file. This is an excellent file to help you take advantage of waiting time. The following trick will work for the newspaper but don't try it until everyone else has seen it. Go through the paper, front to back, and circle the articles you want to read. Then go back and read the circled ones. You will find time savings.

8. Buy speed listening equipment. Earlier we discussed getting cassette tapes and listening to them. The average brain can process information faster than most people can talk. Speed listening tape players are available that will speed up the tapes by skipping very short segments and playing the rest of the tape at normal speeds. The pitch of the speech and music is normal but the elapsed time is reduced to half the normal time.

9. Know and drill yourself on doing jobs to an appropriate level of quality. For example, a punch press tool designed for millions of pieces will

be made to a different quality standard than a temporary tool to make 15 pieces. Some of your projects need to be done `quick and dirty` and others need excellence. Know the difference. Inappropriate quality is a time killer.

10. Homework. The best way to get the highest return on investment from meetings that you run is to insist that people do their homework. The least efficient meeting is one where people sit around and watch each other think! Always schedule preparation time for meetings that you attend.

11. If you feel overwhelmed with different projects, sit quietly for one or two minutes and allow worries to surface. List the worries that occur to you. The worries that occur first might be stopping you from the rest of your work. If possible, try to put the worry list to bed first.

12. After the worries are handled, decide what are your most important jobs, which is setting priorities. Do your highest priority, or most anxiety-producing items, early in the day.

13. Always work to complete what you start. Going back will always cost you time. Experts even extend this admonition to reading the mail. Never pick-up a piece of mail and put it down to deal with later. Pick it up once and deal with it. This applies to E-mail also: trash it, file it, act on it, or forward it!

14. Gain efficiency by grouping related activities together. For example, make all telephone calls together or assign all estimates to MWO's at the same time.

15. Divide larger projects into sub-projects. This follows the philosophy of Project management scheduling. Give yourself the extra motivation of allowing a completion (of a sub-project) everyday.

16. Gain control of your own projects. Reverse load larger projects. Reverse loading starts with the completion date and works backward to the beginning of the project. This sequence gives you logical sub-projects and milestones to see if you are on schedule.

17. Use polite means to end telephone conversations that aren't going anywhere. They include "Glad you called, I have a meeting and can I call you back," then call the person back at 4:50PM to `chat.'

18. Look at your junk mail. Ask the question, which types of mail are useful. Have yourself removed from the useless lists (prepare a form post-card with "please remove my name from your list, I am no longer a prospect for your solicitations," tape their label to the post card).

Companies love to get these cards because it saves them money on wasted mailings.

Junk E-mail is more complicated. Set your Spam filter to catch as much as you can. Consult with your IT people with regard to complaining about the really bad stuff. Common wisdom is not to click the unsubscribe button (unless you know the firm is legitimate).

19. There are several techniques to shorten meetings:
    - Have an agenda, and circulate the agenda to the attendees
    - Know what you want to accomplish at the meeting
    - Stick to the agenda
    - Schedule the meeting at 4:45pm or at 11:15am in the lunch room.
    - Try not to get invited if the subject doesn't concern you, or your work group.
    - Start and end on time
    - Hold the meeting with everyone standing up (to shorten it).

20. Do it now.

Notes to Chapters

Attributes of World Class Maintenance management
20 Steps to World Class Maintenance was an article first published by the author in Maintenance Technology Magazine in December 1992

Evaluate your maintenance department
Revised several times, but the original questionnaire was prepared by Jay Butler

Maintenance Quality Improvement.
W.E Deming, Out of the Crisis MIT Press

Maintenance crewing
Based in part on Don Nyman's work and Maintenance Planning, Scheduling and Coordination by Don Nyman and the author

Managing maintenance through planning and scheduling
Maintenance Planning, Scheduling and Coordination by Don Nyman and the author

Maintenance Work Order
Additional status codes from Don Nyman in his seminar text, Maintenance Management)

Maintenance Management by Jay Butler published by University Seminar Center, now Saddle Island in Boston.

RCM and PMO (Reliability Centered Maintenance, PM Optimization)
Reliability-centered Maintenance by John Moubray published by Industrial Press (address in resource section). This is one of the best texts by one of the leading thinkers in this area.

PMO materials borrowed from the work of the developer of the concepts Steve Turner

PM (Preventive Maintenance)
For more detailed information consult The Complete Guide to Preventive and Predictive Maintenance by the author

TPM (Total Productive Maintenance)
We would like to acknowledge the ground breaking work of Nakajima and Suzuki. Much of the information on TPM in this section is derived from the writings of Seiichi Nakajima (Introduction to TPM, and TPM Development Program as published by Productivity Press.

Information about the originator of TPM (excerpted from an article titled "Lessons from the Guru's" published in Industry Week August 6, 1990.

Other articles and book on TPM used for background and not specifically mentioned:

TPM, An American Approach by Terry Wireman published by Industrial Press Inc, NY. Mr. Wireman is one of America's leading authorities on TPM and in this book he has adapted the Japanese ideas to the American situation.

Just Call Him Mr. Productivity', Industry Week, May 21, 1990 by John Sheridan founder of Productivity Press the major publisher of TPM books.
Will the Real TPM Please Stand Up?', Maintenance Technology Magazine January 1991 by Edward Hartman, American TPM Institute.

`TPM: More Alphabet Soup or a Useful Plant Improvement Concept?', Plant Engineering, Feb 4, 1993 by William M. Windle, A.T. Kearney

Predictive Maintenance, Managing condition based maintenance technology

Quotation and statistics from J. B. Humphries in an article in the September 1988 Maintenance Technology titled Analyzing Predictive Maintenance Needs

Infrared: List of information about infrared from promotional material supplied by Hughes Aircraft Company, Probeye Marketing, Carlsbad, Ca.

Stores
Some elements of the formulas were adapted from The Production Managers Handbook of Formulas and Tables by Lewis Zeyher published by Prentice Hall in 1972

Shutdowns, Outages and Project management
Some of the information on the Gannt, CPM, and PERT charting methods was taken from "Production and Inventory Control" by Plossl and Wight published by Prentice Hall in 1967 and the 1990 Scheduling Guide for Program Managers by Defense Systems Management College.

Shutdowns, Turnarounds, and Outages by Mike Brown. Course by New Standard Institute. I present this course for New Standard and learned much of the background from the training materials

Craft training
Larry Davis' leading book on adult training and education Planning, Conducting, and Evaluating Workshops. This excellent text on adult teaching is available from University Associates, Inc. 8517 Production Ave., San Diego, Ca. 92121

Supervisor evaluation clinic
We would like to thank the firms listed for their input in the form of discussion and questionnaires. Some of the firms that participated include (in alphabetical order): Amoco, AT&T, Baldor Electric, Bendix, Betz Laboratories, Bucks County Community College, City Service Oil, Clements Food, CSX, Cumberland Farms, Goodyear Tire, Honeywell Bull, IBM, IKEA, Indian Health Service, Johnson Controls, Kerr-McGee, OK Steel & Wire, Republic Gypsum, Rider College, Ross Laboratories, 3M, Texaco, Town of West Hartford, Trenton Housing Authority, Valley National Bank, Western Farmers Co-op, Wheatly Pump and Valve and Xerox Corp.

IE-NS-TF-PJ The test is designed to measure a person's personality traits in Jungian terms (Carl Jung first described the traits in 1921). Others have worked on the test, including Myers (1962) and Hogan (1979). If you are interested in detailed descriptions of the 16 types write to: B & D Book Company, 1400 W. 13th Sp. 128, Upland, Ca. 91786 for excerpts from Please Understand Me, Character and Temperament Types.

Introduction to Type by Isbel Myers published by Consulting Psychologists Press, Inc. 3803 E. Bayshore Rd., Palo Alto, Ca. 94303. This publisher owns the copyright for the test, which also lists several good references for people who want to peruse the topic.

Delegation Adapted from the National Seminar course called "How to get more done." .
The One Minute Manager by K. Blanchard, PHD and S. Johnson, MD.

Stated by Lee Minor teaching How to Supervise People offered by Fred Prior Seminars.

Communication Questionnaire by Angelina Rodriguez

Time management in the maintenance pressure cooker

Napoleon Hill's ideas for success. Discussed extensively in his books.
Some of the time-killer items were based in part on an article "Doing Time," that appeared in the February, 1987 issue of Microservice Management a magazine for people and companies that service microcomputers.

We can thank Robert Allen and his associates at Challenge Systems for combining the work of Napoleon Hill and others with their own ideas to come up with the seven daily habits of success.

According to Jeffrey Mayer in his book If You Haven't Got Time To Do It Right, When Will You Find Time To Do It Over?

In The Ninety Minute Hour author Jay Levinson shows the high value of time by telling us to get 90 minutes of work out of every hour

Books:

If you haven't got time to do it right, when will you find time to do it over? by Jeffery Mayer published by Simon and Schuster New York. This short snappy book has excellent ideas and attitudes for the harried supervisor.

The Management of Time by James McCay published by Prentice Hall, Englewood Cliffs, New Jersey. This is an older book (first published in 1959) with some very relevant and up-to-date ideas. One of the great concepts is the quotation that starts chapter two from Lynn White, Jr. "We live in an era when rapid change breeds fear, and fear too often congeals us into a rigidity which we mistake for stability." That was from 1959!

The Ninety Minute Hour by Jay Conrad Levinson published by E.P. Dutton, New York. Jay Levinson is an excellent author who, judging from the quality of his work, must practice what he preaches. This book highlights how to squeeze more productivity out of each hour. In my view these ideas are essential to keep up.

The One Minute Manager by Dr's Blanchard and Johnson published by Berkley Books, New York. This is the book that the section in Module 3 was based on it takes about 2 hours to read, and it's worth it! These techniques are very effective and surprisingly easy to follow.

Maintenance Management Books

The Complete Guide to Preventive and Predictive Maintenance by Joel Levitt published by Industrial Press (2001), NY. This book is a complete description of preventive maintenance. Like the book on planning it is highly focused, with worksheets, checklists, etc www.industrialpress.com

The Complete Handbook of Maintenance Management by John E. Heintzelman, published by Prentice-Hall, New Jersey (TS192.H44). This book is a very readable overview of the maintenance field.

Effective Maintenance Management: Risk and Reliability Strategies for Optimizing Performance By V. Narayan (2004) Published by Industrial Press, NY. Book shows why we do what we do, and how to deliver maintenance services effectively.

Handbook of Building Maintenance Management by Mel A. Shear, published by Reston Publishing Co. (A Prentice-Hall Company) located in Reston, Va. 22090 (TH3361.S45). This is a nuts and bolts book of building management. It combines ideas on the management of maintenance with the doing of maintenance. Excellent for maintenance departments that support the facility in addition to the equipment.

Handbook of Maintenance Management by Joel Levitt published by Industrial Press (1997), NY. This is a complete survey of the field. www.industrialpress.com

Housekeeping Handbook for Institutions Business and Industry by Edwin Feldman. This excellent book (almost 500 pages, now quite old and still useful) is available from Ed himself (address in the people section of resources). It is a complete handbook designed for the manager of the housekeeping department.

Internet Guide for Maintenance Management by Joel Levitt published by Industrial Press (1999), NY. Help for supervisors interested in using the Internet to make their jobs easier. www.industrialpress.com

Introduction to TPM by Seiichi Nakajima published by Productivity Press, P.O. Box 3007, Cambridge, MA 02140, 617-497-5146. This book gives a complete overview of TPM. It and its companion volume below are essential reading to understand TPM.

Maintenance Management by Don Nyman available from Saddle Island, 100 State Street, 4th Floor, Boston, MA 02110. This text is a summary of maintenance lore and information from Don's 30 years as a top-level consultant. An excellent questionnaire to evaluate your department.

Maintenance Management and Regulatory Compliance Strategies By Terry Wireman 2003 Industrial Press, NY. Maintenance work is bounded by regulators and regulations. Wireman shows the relationship of good maintenance practices to the regulations we work under.

Maintenance Planning, Scheduling and Coordination by Don Nyman and Joel Levitt published by Industrial Press (2001), NY. This book is a complete description of maintenance planning scheduling and coordination. It is a highly focused book with worksheets, checklists, etc www.industrialpress.com

Managing Maintenance Planning and Scheduling by Mike Brown published by Audel Press (2004). This book is the basis for the popular course on the same name at New Standard Institute. As such it gives good background and justification for planning. Some excellent work sheets and models.

Managing Maintenance Shutdowns and Outages by Joel Levitt published by Industrial Press (2004), NY. This book is a complete step-by step guide to the conception, planning, scheduling and execution of large maintenance jobs, including partial or total plant shutdowns.

The Maintenance Scorecard Creating Strategic Advantage by Daryl Mather published by Industrial Press (2004), NY. This is a new work that rigorously looks at measuring maintenance. It is destined to become one of the standards of the business.

RCM II Reliability-centered Maintenance by John Moubray published by Industrial Press NY. Excellent complete review of the field. RCM will become increasingly important as firms downsize and wonder how to improve reliability.

Total Productive Maintenance, An American Approach by Terry Wireman, published by Industrial Press, 200 Madison Ave. New York, NY 10016. This book explains the steps for TPM for an American organization. It has many excellent ideas for the organizations considering TPM.

TPM Development Program by Seiichi Nakajima published by Productivity Press, P.O. Box 3007, Cambridge, MA 02140, 617-497-5146. This book gives a complete systems design to setting up and day to day working of a TPM system.

Uptime by John Campbell published by Productivity Press, Portland, OR (1995). An excellent strategic look at maintenance options and imperatives.